From Fields to Fortunes

The Strategy Behind Africa Feeding Itself and the World

By Godwin Kojo Ayenor, PhD

Black Seeds Publishing
Clarksdale, MS 38614
www.blackseedspublishing.com

Ordering Information:
Quantity sales. Special discounts are available on quantity purchases by corporations, associations, and others. For details, contact the publisher at the address above.

Printed in the United States of America

ISBN-13 (paperback): 979-8-9878956-5-8
ISBN-13 (hardback): 979-8-9878956-4-1

"I've spent my career working in African agriculture and the international development landscape because I believe—deeply—that our farmers can prosper, our rural communities can thrive, and our continent can feed itself and the world. This book is my contribution to making that vision a reality."

— Dr Godwin Kojo Ayenor
(Founder & CEO of STRED Consult Ltd)

TABLE OF CONTENTS

From Data to Decisions

Good policy is not paperwork; it is change you can see. STRED helps partners design, evaluate, and improve programs that actually work.

- Monitoring and evaluation frameworks
- Baselines, endlines, learning briefs
- Decision-ready reports for action

www.stredconsult.com

Prologue

For thirty years, I have carried one message into farms and policy rooms: Africa's agriculture is more than our gold mine. We've been digging in the wrong places. This book is a call—and a practical, adaptable roadmap—to move the shovel toward the practices, institutions, and markets that turn potential into prosperity.

Let me tell you about Kwame.

I met him in 2003 on a cocoa farm in Ghana's Eastern Region. Bright man. The kind who never missed a training session. He could recite the principles of integrated pest management and organic farming practices better than some extension officers I know.

But one morning, I found him spraying synthetic pesticides on his organic-certified cocoa.

"Godwin," he said—no drama, no apology—"my children's school fees are due next week. These capsid bugs will destroy my harvest. What do you want me to do—let my children sit at home while I wait for neem extract to work?"

In that moment, I learned what many experts still refuse to learn: Kwame wasn't ignorant. He was surviving.

Kwame's dilemma wears different uniforms across Africa—cocoa in Ghana, maize in Kenya, rice in Nigeria, cassava in Côte d'Ivoire —but the story is the same. A farmer can know the "right" practice and still choose the "wrong" one, not because he is backward, but because the system has made survival more urgent than sustainability. When your harvest is your bank account, your children's school fees are a deadline, not a theory.

And yet, in far too many meetings across this continent, we keep returning to the same question, dressed up in different PowerPoint slides:

Why don't farmers adopt our recommendations?

That question is not innocent. It is a worldview. It assumes knowledge lives on one side—research stations, ministries, donor reports—and that farming communities are empty vessels to be filled. It turns extension into a delivery truck, and farmers into "end-users." Once you accept that framing, the solutions become predictable: more training sessions, more inputs, more campaigns, more targets, more "sensitisation." And when adoption stays low, we blame the farmer, then design the next project to try again.

But what if the problem is *not* the farmer?

What if the problem is the *model*?

Here is what keeps me up at night—and what gets me out of bed every morning with fire in my belly: Africa has the land, the labour, the intelligence, and the markets to feed itself with dignity and to export value, not just raw commodities. Our young people are not the problem; they are the greatest agricultural asset we have. Our farmers are not waiting for salvation; they are waiting for systems that respect their reality.

So why do we still import so much of what we can grow? Why do the people who feed this continent sometimes struggle to feed their own families?

Because we've been asking the wrong questions.

The better question is not, *"How do we make farmers adopt?"* The better question is: *"How do we build a system where farmers can afford to choose well—and where knowledge is co-created, not delivered?"*

That is the journey this book begins. And cocoa—Kwame's cocoa —will be one of our anchors. Because inside that one farm is the whole African story: science and tradition, policy and poverty, markets and moral choices, survival and the possibility of prosperity.

When the Rains Come, Everyone Knows

When the rain is coming, everyone knows. You don't need a weatherman to tell you it's raining. You feel it.

Right now, I can *feel* the rains coming for African agriculture. And I'm not talking about climate change (though that's real too). I'm talking about a convergence—a moment when everything that needs to happen is finally possible.

Mobile phones in the hands of farmers who never had extension officers visit them. Climate-smart techniques that work with our weather, not against it. Regional trade opening up markets that colonialism fragmented. Young entrepreneurs building agribusinesses that rival anything in Europe or America.

But here's the thing about rain: if your roof has holes, you'll still get wet.

I've watched brilliant research only partially utilised or completely gathering dust in filing cabinets (see references). Typically, I've seen well-meaning policies create disasters because nobody asked the farmers what they actually needed. I've also witnessed millions of dollars poured into projects that collapsed the moment the donors left.

The gap between what we know and what we do—that's one of the main barriers standing between Africa and agricultural transformation.

From the Lab to the Land

My journey started in the fields. As a native of Nsawkaw, but born in Wenchi, Brong-Ahafo Region, farmer's blood runs through me. Also, though my late father was a Police Commissioner in the 1970s, he always saw himself as a farmer. Hailing from Asitey near Odumase Krobo in the Eastern region, he had farms wherever he worked across Ghana.

But life took me to Cuba from my secondary through pre-university and university education, then back to Ghana, by way of British Aid projects and TechnoServe sponsored by the US government. Then to Wageningen University in the Netherlands where I defended my PhD.

I've worked with the World Bank, USAID, WFP, FAO, UNDP, ILO, EU, GIZ, USDOL and others. I've had the opportunity to advise the Cabinet and Ghana's government on agricultural policy. I've evaluated projects worth millions of dollars in Ghana and across West Africa.

But you want to know where I learned the most? *In the farms and communities.* Sitting under mango, cocoa, cashew, or baobab trees with farmers, listening to them explain why our "perfect" solutions don't work in their reality.

That's why I founded STRED Consult (www.stredconsult.com). Because I got tired of the gap between theory and practice; oh yes, and the immutable top-down technology transfers... Tired of watching consultants fly in, write reports, and fly out—leaving farmers exactly where they found them.

I've learned something in these thirty years though: transformation doesn't come from grand strategies written in air-conditioned offices... it comes from understanding the dance between policy and practice, between global trends and local realities, between what our ancestors knew and what technology can do.

What This Book Will Give You

Whether you're a policymaker trying to feed your nation, an investor looking for the next frontier, an entrepreneur building an agribusiness, or a development worker tired of projects that don't work—this book will give you something most agricultural books don't: **a roadmap based on what actually works.**

There are programmes that are theoretically sound, but fail abysmally in practise. I call them "more of the same failures presented differently as new".

This book will not overly elaborate on Africa's agricultural potential; we all know that. Instead, it offers the strategy - call it - the adaptable blueprint.

Think of this book like a good *groundnut soup*. You need the right ingredients, yes. But you also need to know when to add them, how long to stir, when to add water, and when to let it simmer. The recipe alone won't teach you that—you need someone who's cooked it before, burned it a few times, and finally got it right.

Here's what you'll discover:

- Why the old development model (anchored on the lineal perspective) is yesterday's dead crop —and here's the next harvest (grounded on systems thinking within the constructivist perspective):

- The five systems that must work together for transformation (like fingers on a hand—separate, they're weak; together, they're a fist)

- How mobile technology is leapfrogging infrastructure—real examples from Ghana, Nigeria, Kenya

- The policy innovations that actually move the needle—and why most agricultural policies fail before they start

- How to build value chains that lift farmers up instead of squeezing them dry

- Climate-smart practices that increase yields while protecting our land—not theory, but proven methods

- Why organic and sustainable farming isn't just good ethics—it's good business (i.e., Organic cocoa in Ghana)

- How to bridge the research-to-practice gap that has frustrated us for decades

Each chapter mixes rigorous analysis with the author's real stories and experiences—showing you not just *what* should happen, but *how* to make it happen in the messy, resource-constrained world where most of us actually work.

The Fortune in the Fields

"From Fields to Fortunes"—I chose this title carefully. Like how we choose the right cutlass for clearing bush.

For too long, we've treated African agriculture like a charity case. A problem needing aid. A sick patient needing medicine. *"Daabi!" (Meaning no in Akan language);* It's time we see it for what it truly is: **the greatest economic opportunity of the 21st century.**

When I talk about fortunes, I'm not just talking about money (though the money is real and it's big). I'm talking about:

- **Food security fortunes**—nations that feed themselves and export to the world

- **Environmental fortunes**—farming that heals the land instead of killing it *(not the false fortune of illegal mining)*

- **Social fortunes**—villages where young people see a future worth staying for

- **Knowledge fortunes**—our ancestors' wisdom meeting modern science (including converging beta & gamma sciences)

- **Dignity fortunes**—farmers who are entrepreneurs, not beggars (part of a structured trading system - not price takers)

These fortunes are within reach. I've practised them and seen them work. I've touched them. I've helped create them (i.e., maize triangle and organic cocoa belt in Ghana).

But only if we're willing to challenge the old ways, embrace complexity, and do the hard work of implementation.

How to Use This Book

You don't have to read this book from cover to cover like a novel (though I hope you will). Each chapter stands alone—like different dishes at a party. Take what you need when you need it.

Throughout, you'll find:

- **Digital Innovation Spotlights**—QR codes linking to the other half of the story: This book builds the future. Its companion names what is destroying it. Scan the QR code to get the full picture

- **Implementation Frameworks**—practical tools you can adapt to your situation

- **Reflection Questions**—to help you apply these insights to your own challenges

- **Resource Libraries**—curated tools, templates, links to over 30 International development reports co-authored or authored by yours truly, and further reading

And because technology and innovation are driving changes in agriculture faster than the weather these days, this book connects to a living ecosystem of resources at **stredconsult.com** —updated case studies, community discussions, and tools to support your journey.

An Invitation

Remember Kwame, the cocoa farmer I told you about at Brong-Densuso? He eventually became one of our most successful organic cocoa producers. Not because we gave him more training. Not because we scolded him for using pesticides.

We helped him access a value chain that rewarded his sustainable practises with premium prices. We connected him with other farmers facing the same struggles. We made the right choice, the *profitable* choice.

His transformation wasn't about knowledge alone. It was about systems, converging ideas and sciences, incentives, and support structures including new ways of thinking and training that worked within his reality, not against it.

That's what this book is about. Not wishful thinking. Not theoretical frameworks that look good on paper but collapse in the field. **Practical, proven pathways from where we are to where we need to be.**

Africa's agricultural transformation isn't coming. *It's already here.* Like rain clouds gathering on the horizon. The question is whether you'll prepare your roof—or watch from inside while others harvest the blessing.

More of the same will not work. New answers are required to address today's challenges.

Let's do something new together.

"Kojo, this is an impressive and deeply grounded manuscript — the clarity, the lived experience, and the systems framing are all strong. Finally, your credibility gives you the license to call the political economy forces out. That sits beneath many of the failures you diagnose. You describe institutional gaps, policy incoherence, and implementation breakdowns, and these are rarely neutral accidents. They are the logical outcomes of incentive structures, patronage networks, and political bargains that shape how agricultural systems function.

In short, the manuscript is already strong and adding the political economy angle made it definitive. Thanks for giving me the opportunity to review this manuscript and making my inputs."

—Mr Yaw Nsarkoh
Strategic Advisor, Professional Coach & Director at University of Cambridge, London UK

Africa's Agricultural Potential - Untapped and Underestimated

You know what makes me laugh—and cry at the same time? A recent report found that Africa accounts for 60% of the world's uncultivated arable land. Sixty percent! That's like having a room full of gold and begging for coins on the street.

But here's the painful truth: we spend an estimated $78 billion every year importing food. Some countries, such as Zimbabwe, Guinea, and Sudan, spend more than 100% of their annual foreign-currency receipts just to feed their people. *Eii, hmmm!* How did we get here?

Let me tell you about another conversation I had with a farmer in the Northern Region of Ghana. His name is Alhassan. He farms maize on five acres of land his grandfather left him. Good land.

Rich soil. But when I asked him how much he harvests per acre, he looked at me and said, "Dr Ayenor, I harvest what God gives me."

Now, I respect faith. But God also gave us wisdom to make choices, didn't He? This is a systems failure - faith must be backed with works.

Alhassan's yield was about 1.2 metric tons per hectare. Meanwhile, farmers in Brazil—with similar conditions—are getting 5 to 6 metric tons per hectare. The difference isn't the land. It's not even the rain. It's the *system*.

The Numbers Don't Lie: Africa's Agricultural Paradox

Let me give you the facts, plain and simple. What we have:

- I repeat, 60% of the world's uncultivated arable land (see References section at the end of the book)

- **Agriculture employs more Africans than any other sector**—about 52% of employed people in Sub-Saharan Africa work in agriculture

- Our agricultural market could grow from $280 billion in 2023 to $1 trillion by 2030

- A young, growing population—the median age in Africa is 19 years old

What we're doing:

- Importing $78 billion worth of food annually

- 794.7 million people—nearly 60% of our population—face food insecurity

- The lowest agricultural productivity per worker in the world

- Only 2.1% of public budgets go to agriculture—far below the 10% target we set for ourselves in the Malabo Declaration

You see the problem? It's like having a Ferrari in your garage but walking to work every day because you don't know how to drive it.

Why Are We Here? Understanding the Roots

1. The Colonial Hangover

Let me be honest with you. We can't talk about African agriculture without talking about colonialism. I'm not one to dwell on the past, but if you don't understand history, you'll keep repeating it. Your mistakes.

Before colonialism, our ancestors grew what they needed: yams, millet, sorghum, and cassava. They understood the land. They worked with the seasons. Then the Europeans came and said, "No, no, no. Grow cocoa. Grow coffee. Grow cotton. We need these in Europe."

So we did. And we're still doing it.

Today, 14.8% of Côte d'Ivoire's land is used for cocoa production. Côte d'Ivoire produces 40% of the world's cocoa. But you know what? The 5 million people—20% of their population—who depend on cocoa farming remain in chronic poverty. Familiar story, right? It is no different in Ghana.

Why? Because we export raw cocoa beans for pennies, and Europe turns them into chocolate bars worth dollars. We do the hard work. They take the profit. *Ɛyɛ ya* (It's painful).

2. The Green Revolution That Passed Us By

In the 1950s and 1960s, Asia and Latin America experienced what they called the "Green Revolution." New seeds. Better fertilisers. Irrigation systems. Agricultural productivity exploded.

Africa? We were left behind.

Why? Some say it was because our soils are different. Our climates are more diverse. Our farms are smaller. All true. But the real reason is simpler: **nobody invested in us**. And when they did invest, they invested in export crops, not in feeding our own people.

From my research in Ghana's cocoa sector, I saw this firsthand. The Cocoa Research Institute of Ghana (CRIG) developed technologies that could increase yields from 400 kg per hectare to 1-5 metric tons per hectare. But less than 4% of farmers adopted these technologies.

Why? Because the technologies were developed *for* farmers, not *with* farmers. Nobody asked Kwame or Alhassan what they actually needed, nor did they work with them to find a common pathway.

3. The Smallholder Trap

About 80% of Africa's food is produced by smallholder farmers. These are farmers with 1 to 5 hectares of land. They use traditional methods. They can't afford fertilisers. They don't have access to credit. They can't afford to take risks.

It's like asking someone to run a marathon in flip-flops. They might finish, but they'll never win.

And here's the thing: we keep treating smallholder farming as a "development problem" instead of recognising it as a *business opportunity*.

4. Land Tenure Insecurity

Here's what will shock you. Between 2005 and 2017, large-scale land deals in Africa, mainly involving foreign nations and companies, totalled about 22 million hectares. I am not too sure whether Africans had shares in them.

In some cases, these companies come and say, "We'll develop your land. We'll create jobs. We'll bring technology." But what really happens? They grow crops for export— often some feedstock that was never processed into biofuels —while local communities lose access to land they've farmed for generations.

In Ghana, I've seen communities displaced by large plantations. Companies promise jobs, but they bring in their own workers. They promise development, but the roads they build only lead to their farms, not to the villages. Just like during the colonial days.

And because most African farmers don't have formal land titles, they have no legal protection. One day, you're farming your grandfather's land. The next day, a company shows up with a lease agreement signed by someone in the capital city who's never even seen your village. This is similar to how some illegal mining operations in Ghana, began. Some were equally given an exploration or mining concession from the Capital, without community involvement.

The Potential: What We Could Be

Now, let me paint you a different picture. Because despite all these challenges, I'm not a pessimist. I'm a realist with hope.

The $1 Trillion Opportunity

According to the African Development Bank, Africa's food and agriculture market could grow from $280 billion in 2023 to $1 trillion by 2030. That's not a dream. That's a projection based on current trends—population growth, urbanisation, rising incomes, and changing diets.

But here's the key: most of that $1 trillion will be spent on food imports unless we act now.

The Youth Dividend

Africa has the youngest population in the world. By 2030, we'll have more young people entering the workforce than any other continent.

Right now, young people are running away from farming. They see it as old-fashioned. Backbreaking. Unprofitable. And honestly? They're not wrong—*if farming stays the way it is*.

But what if we could show them a different kind of farming? Farming with drones. Farming with mobile apps that tell you when to plant and when to harvest. Farming that connects you directly to buyers in Accra, Lagos, or Nairobi. Farming that makes you an entrepreneur, not a peasant.

That's the future we can build.

The Climate Advantage

Yes, climate change is real. Yes, it's affecting us. But Africa also has advantages.

We have year-round growing seasons in many regions. We have diverse agro-ecological zones—from the Sahel to the rainforests to the highlands. We have crops that are naturally drought-resistant and heat-tolerant.

And more recently, the emerging agro-technological solutions from quantum nano sciences appear to significantly address many yield-limiting factors.

While Europe and North America worry about their crops failing due to extreme weather, we can make it a strategic priority to become the world's most climate-resilient agricultural producer.

But only if we invest in it. Enough about empty promises and endless debates, politics and talk shops. Let's act now!

Real Examples: It's Already Happening

Let me give you some hope. Because transformation isn't just theory—it's already happening in pockets across the continent.

Rwanda: the Policy Success Story

Regarding the Malabo Declaration targets, Rwanda has been among Africa's top Malabo performers, repeatedly rated on track in earlier review cycles. How did they do it?

Simple: **political will.**

The government made agriculture a priority. They invested in infrastructure. They supported farmer cooperatives. They created policies that work for smallholders.

Result? Agricultural productivity increased. Poverty decreased. Food security improved.

Kenya: the Mobile Revolution

In Kenya, a platform called M-Farm connects farmers directly to buyers using mobile phones. Farmers can check market prices, find buyers, and negotiate better deals—all from their phones.

No guessing what price to sell at. Just direct, more transparent transactions with farmers, who are better able to compare prices and negotiate. Where they refuse to remain price takers and take their place as key actors of the food supply chain.

This is what I mean by leapfrogging. We don't need to build the same infrastructure Europe built over 200 years. We can jump straight through digital solutions.

Ghana: the Cocoa Turnaround

In my own work in Ghana, I saw what happens when you give farmers better prices and better support. They respond appropriately and, in many cases, show innovation.

In 2002, Ghana's cocoa production was about 340,600 metric tons. Farmers were getting only 40% of the Free on Board (FOB) price. Many were abandoning their farms.

Then the government began to increase the farmer share towards a policy target of 70% of the FOB price.

By 2004, production jumped to 736,911 metric tons—more than double!

The lesson? **Farmers respond to incentives.** Give them a fair price, and they'll produce. Give them support, and they'll innovate.

The Path Forward: Five Pillars of Transformation

Based on my 30 years of experience—from the field to the policy table —here's what we need to do:

1. Invest in Smallholders, Not Just Large Farms

Smallholder farmers aren't the problem. They're the solution—if we support them properly. This means:

- Access to credit at reasonable rates
- Access to inputs (seeds, fertilisers, tools) at affordable prices
- Access to markets through cooperatives and digital platforms
- Access to knowledge through extension services that work, including digital platforms and/or connecting them to agribusiness firms (medium to large-scale as outgrowers).

2. Secure Land Rights

No farmer will invest in improving their land if they don't know whether they'll still own it next year or at least have medium- to long-term access. We need:

- Clear, documented land titles
- Protection against land grabs
- Support for communal land management systems that respect traditional ownership while providing security

3. Build Infrastructure

You can grow the best tomatoes in the world, but if they rot on the way to market because the road is impassable, what's the point? We need:

- Rural roads that connect farms to markets
- Storage facilities (i.e., cold chains) to reduce post-harvest losses (currently 30-40% in many countries)
- Processing facilities so we can add value before export
- Irrigation systems, so we're not entirely dependent on rainfall (or even more cost-effective means of hydrating crops)

4. Embrace Technology

The future of African agriculture is digital, and the tools already exist to improve decisions, reduce losses, and increase productivity. This includes:

- Mobile phones and practical farm apps
- Satellite imagery and weather forecasting tools
- Drones and emerging precision technologies
- Training and local digital platforms that farmers can actually use at scale

These aren't luxuries. They're necessities. And the good news? Most African farmers already have mobile phones. We just need to give them the right apps and training (i.e., www.stredconsult.com).

5. Change the Narrative

We need to stop seeing agriculture as a "poor man's job" and start seeing it as a business. This means:

- Teaching agribusiness in schools, not just farming
- Celebrating successful farmers as entrepreneurs

- Creating pathways for young people to enter agriculture as a career, not a last resort

The Choice Is Ours

Let me end with this. Africa has everything it needs to feed itself and feed the world. We have the land. We have the people. We have the climate. We have the potential, and this book provides an adaptable roadmap.

What we need is the *will*.

Once there is a will, there is always a way. When people come together and decide to do something, they can achieve it.

The question is: Do we want it badly enough?

Because the alternative is clear. If we don't transform our agriculture, we'll keep spending billions importing food while our youth migrate to cities looking for jobs that don't exist. We'll keep exporting raw materials while others capture the value. We'll keep talking about potential while others eat our lunch—literally. And our economic transformation goals will only remain as dreams.

But if we act now—if we invest, if we innovate, if we empower our farmers—then by 2030, Africa won't just be feeding itself. We'll be feeding the world.

And that $1 trillion opportunity? It will be ours.

The choice is ours. The time is now.

Yɛ bɛ tumi! (We can do it!)

DIGITAL INNOVATION SPOTLIGHT

https://stredconsult.com/land-and-legacy-books/

REFLECTION QUESTIONS

1. What is your country's current agricultural import bill? How much could it be reduced by increasing domestic production?

2. What are the top three crops your country imports that could be grown locally?

3. Who are the key stakeholders in your agricultural sector? Are smallholder farmers at the table when decisions are made?

4. What is one policy change that could immediately improve agricultural productivity in your context?

5. How can you personally contribute to transforming agriculture in your community or country?

Chapter 2

The Barriers We've Built Ourselves

"The problem is not that we don't know what to do. The problem is that we keep stopping ourselves from doing it."

Let me share more about some of my conversations with another cocoa farmer at Brong-Densuso. His name was Obutey, and he had been farming cocoa for over thirty years. When I asked him why he wasn't using the improved pest control methods that our research had proven to work—methods that could double his yields—he looked at me with tired eyes and said, "Godwin, I know it works. But who will buy my organic cocoa? COCOBOD only buys what they can spray." That was the status quo, and the rest is history.

However, that conversation remains relevant because it captures the essence of Africa's agricultural paradox: we have the knowledge, we have the land, we have the people—but we've built a system that makes it nearly impossible for farmers like Obutey to succeed due to removable market failures.

The Colonial Hangover

When the Portuguese arrived on our shores in the fifteenth century, they did not meet an empty continent. They met communities that farmed, traded, stored, and adapted—systems shaped by ecology, seasons, and hard-earned knowledge.

Then came Europe's chartered trading companies. They didn't come to learn from us. They came to extract. Over time, incentives shifted: land, labour, and policy began to serve distant industries—cotton for textile mills, cocoa for chocolate, coffee for breakfast tables—often at the expense of the diversity that had fed African families for generations.

Independence should have changed this. But here is the uncomfortable truth: **we never truly decolonised our agriculture**.

Today, Ghana still spends billions annually on food imports.

So, we still export raw commodities and import finished products. We ship cocoa beans and buy chocolate. We grow cotton and import clothes. We remain—too often—*hewers of wood and drawers of water* on our own land.

The Policy Trap

I've spent years working with the Ministry of Food and Agriculture, the World Bank, FAO, and countless other institutions. I've seen brilliant policies drafted, ambitious strategies developed, and comprehensive frameworks designed. And I've watched most of them fail, or at best, with extremely limited success. Why?

Because our agricultural policies are often written in air-conditioned offices in Accra or Abuja, by people who haven't stood in a muddy field at 6 – 7 a.m. watching a farmer try to decide whether to buy fertiliser or pay school fees. To sell his maize and pay school fees now, and if not, where and how to keep it for a better price later.

Take Ghana's COCOBOD, for example. It's an institution born from good intentions—to maintain the quality of our cocoa and ensure farmers get fair prices. But in practice, it has sometimes become a barrier to innovation. When we tried to develop an organic cocoa value chain in the early 2000s, we ran into a wall. Initially, COCOBOD's mass spraying programme with synthetic pesticides made organic certification nearly impossible. The farmers wanted to go organic—there was a premium market waiting—but the system wouldn't allow it.

This is what I call "the policy trap": regulations designed to protect farmers that end up imprisoning them. It had to take more than acting as an independent action researcher; it also meant acting as a Ghanaian, helping the farmers collectively reclaim their decision to spray or otherwise, which truly was their right anyway.

This story is part of the many efforts to protect the organic cocoa fields near Akwadum and Koforidua in the Eastern Region.

There is a harder version of this diagnosis that must be named. The barriers described in this chapter — colonial commodity structures, policy traps, underfunded extension, land tenure insecurity — are not simply the residue of neglect or resource scarcity. Many of them persist because someone benefits from their persistence. Input subsidy systems that are poorly targeted but widely announced serve political visibility, not farmer productivity. Land tenure ambiguity that locks smallholders in insecure arrangements also locks in the power of those who control access to land. Extension services too underfunded to reach farmers cannot threaten the intermediaries who profit from information asymmetry.

This is not cynicism. It is a framework for understanding why good ideas fail. Good intentions aren't enough. If your programme succeeds, someone loses money, power, or relevance. If you never figured out who that is — and never planned for their pushback — they will kill your programme, and you won't even see it coming.

The Infrastructure We Never Built

Let me paint a picture for you. A tomato farmer in northern Ghana harvests her crop. The nearest big market is 50 kilometres away. The road is unpaved, full of potholes, and impassable during the rainy season. She has no cold storage. No processing facility nearby. By the time her tomatoes reach the market—if they reach the market—30% have spoiled.

Now imagine the same farmer in the Netherlands. She harvests her tomatoes. A refrigerated truck picks them up within hours. They're processed, packaged, and on supermarket shelves across Europe within hours or maximum days. Minimal waste. Maximum value.

The difference isn't the farmer's skill or the quality of the tomatoes. It's infrastructure.

Poor road, rail, and harbor infrastructure adds 30-40% to the costs of goods traded among African countries. Think about that. Before a farmer even negotiates a price, she's already lost nearly half her potential profit to bad roads and non-existent storage facilities.

And it's not just physical infrastructure. It's digital infrastructure too. In rural Africa, less than 30% of adults have access to the internet, and most of that is 2G or 3G—too slow for the agricultural apps and platforms that could transform their businesses.

The Trust Deficit

During my doctoral research on organic cocoa production, I encountered something that no technical training could fix: a complete breakdown of trust.

The farmers didn't trust COCOBOD. COCOBOD didn't trust the farmers. The organic certification company didn't trust the farmers' association. Some of the farmers' association leaders were more interested in their monthly allowances as government-sponsored spraying workers than in building a sustainable organisation to support organic cocoa farming.

As one farmer told me, "If we openly say we are not members of the Traditional Organic Farmers Association (TOFA), as it was previously named, then our farms would not be sprayed with the Neem extract (ANSE) for free. So, we just keep quiet and look on. But we have no time to attend TOFA's meetings."

This trust deficit is one of the most insidious barriers we faced. It's invisible on balance sheets and policy documents, development reports or even scientific journals, but it's as real as a collapsed bridge. When farmers don't trust the system, they don't invest. When they don't invest, productivity stagnates.

When productivity stagnates, poverty persists. It seems a replay of Farmer-Based Organisations (FBOs), or Cooperative stories that development practitioners are too familiar with. They can also be fixed.

The Research-Reality Gap

I've worked with the Cocoa Research Institute of Ghana (CRIG) for years. They're brilliant scientists doing world-class research. They've developed pest control methods that work. They've bred disease-resistant cocoa varieties. They've created farming techniques that could increase yields.

But here's the problem: *less than 4% of farmers adopted their recommendations*.

Why? Because the research is done *for* farmers, not *with* them. Scientists identify problems in laboratories. They test solutions on research stations. Then, at best, they hand over a manual to train extension officers and expect farmers to adopt them.

This is because farming isn't a laboratory. A technique that works perfectly on a research station with reliable water, quality inputs, and trained staff might fail completely on a small farm with erratic rainfall, expensive fertiliser, and a farmer juggling three other jobs to make ends meet. We hope the research station will extend and then farmers as end users' model in deploying technologies has changed by now.

This is what my thesis was about—bringing farmers into the research process from day one. When we did that with capsid control in organic cocoa, adoption rates soared. Not because the technology was different from the research station and better, but because farmers owned it.

They helped design it.

They understood it.

They trusted it.

"Governments do not put their money where their mouth is. They come out with eye-catching themes to grab attention, but nothing more. Agriculture is treated with a cosmetic approach — and that must change. *From Fields to Fortunes* names this plainly and demands better."

— Dr Emmanuel K. Foe, Agribusiness Expert

The Money Problem

Let's talk about money. Or rather, the lack of it. African governments committed through the Comprehensive African Agricultural Development Programme (CAADP) to spend at least 10% of their budgets on agriculture (i.e., Malabo Declaration). In 2021, the average was 4.1%. Some countries spend less than 1%.

Meanwhile, the World Bank—once a major funder of African agriculture—reduced its agricultural lending from 39% in 1978 to just 7% in 2000.

The message is clear: agriculture isn't a priority.

But it gets worse. When funding does come, it often comes with strings attached. Structural adjustment programmes in the 1980s and 1990s forced African countries to cut agricultural subsidies, privatise extension services, and open markets to cheap imports. The result? Local farmers couldn't compete. Food security collapsed. Rural poverty deepened; it became generational.

And here's the cruel irony: while we were told to eliminate subsidies, European and American farmers were receiving billions in government support. We were asked to fight with one hand tied behind our backs.

The Land Question

At Brong-Densuso, I met farmers who didn't own the land they worked on. They operated under the *abusa* system—a sharecropping arrangement where the landowner takes two-thirds of the harvest and the farmer keeps one-third. Or the *abunu* system, where they split it 50-50.

Now, if you don't own the land, why would you invest in improving it? Why would you build terraces, install irrigation, or plant trees that take years to mature? You wouldn't. You'd extract what you can and move on.

Insecure land tenure is one of the biggest barriers to agricultural investment in Africa. It's not just about ownership—it's about the confidence to invest in the future.

The Gender Blind Spot

Women contribute substantially in food production and processing (60-80%) in Africa. But they own less than 15% of the land. They have less access to credit, less access to extension services, and less access to markets.

I've seen this firsthand. In almost every project I've worked on, the women are there—planting, weeding, harvesting, processing. But when it's time to make decisions or access resources, suddenly they're invisible.

This isn't just unfair. Study after study shows that when women have equal access to resources, yields increase by 20-30%. We're leaving billions of dollars on the table because of some of our outdated gender norms.

The Climate Reality

And then there's climate change, the barrier we didn't build but must now overcome. Rainfall patterns are shifting. Droughts are longer. Floods are more severe. Making predictability elusive. Pests and diseases are spreading to new areas. The agricultural calendar that farmers have relied on for generations is no longer reliable.

In northern Ghana, farmers told me they used to know exactly when the rains would come. Now, they plant and pray. Sometimes the rain comes early and wash away the seeds. Sometimes they come late and the crops wither. Sometimes they don't come at all.

Climate change isn't a future threat. It's a present reality. And it's hitting African farmers—who contributed least to the problem—the hardest. No more lamentations, we need to do something.

Breaking Free

Here's one of those things that keeps me up at night: every single barrier I've described is solvable. We know how to build roads. We know how to reform policies. We know how to secure land tenure. We know how to include women. We know how to adapt to climate change.

The question isn't *what* to do. The question is *will* we do it?

Because here's the truth: these barriers aren't accidents. They're not natural disasters. They're choices. Choices made by governments that prioritise other sectors. Choices made by donors who impose conditions. Choices made by institutions that resist change. Choices made by societies that marginalise women.

And if they are choices, then they can certainly be changed.

In the next chapter, I'll show you what happens when we make different choices. When we put farmers at the center. When we invest in infrastructure. When we reform policies. When we build trust.

I'll show you the convergence—where science meets practice, where policy meets reality, where potential becomes prosperity.

But first, we had to understand the barriers. Because you can't break free from chains you don't see or understand how you got trapped.

Youth and rural opportunity

When youth see a path, illegal mining loses its power. STRED supports training, enterprise models, and systems that make agribusiness a real option.

- Youth enterprise support models
- Agribusiness training and coaching
- Market systems and value chain planning

www.stredconsult.com

Youth belong in the future of farming >

What's Actually Working (And Why We're Not Scaling It)

When I visit cocoa farms in Ghana's Eastern Region, I often ask farmers a simple question: "What's working for you?" Not what the government says should work. Not what the research station recommends. What's *actually* working in their fields, with their resources, under their conditions.

The answers always surprise me. And they should surprise you too.

Because while we spend billions on agricultural programmes that look good on paper, some of the most effective solutions are hiding in plain sight—being practised by farmers, tested by communities, and proven by results. The problem isn't that we don't have solutions. It's that we're not paying attention to the right ones.

Let me show you what I mean.

The LARC Revolution: When Farmers Become Scientists

At Brong-Densuso, a small cocoa-growing community, we tried something different. Instead of telling farmers what to do, we asked them to become researchers.

We established what's called a Local Agricultural Research Committee (LARC)—a group of farmers selected by their community to test solutions to their biggest problem: capsid pests that were destroying their cocoa. These weren't just any farmers. They were representatives chosen by their neighbours, accountable to their community, and motivated by a shared goal: producing organic cocoa for premium prices.

Here's what happened: The LARC farmers learned to scout for pests, conduct experiments, and analyse results. They tested three methods: neem extracts, pheromone traps, and weaver ants as natural predators. Then—and this is the crucial part—they reported back to their community regularly, sharing what they learned.

The results? Farmers who participated in or were exposed to the LARC approach gained complex knowledge about pest ecology that traditional extension services had failed to deliver for decades. They understood *why* certain methods worked, not just *what* to spray. They could make their own decisions based on observation, not just follow instructions.

The lesson: When you treat farmers as partners in research rather than recipients of advice, adoption rates soar. The LARC approach proved more effective than conventional Farmer Field Schools because it deliberately built in community learning and accountability.

Mobile Technology: Putting Extension in Every Pocket

Fast forward to today. Across Africa, digital solutions are driving meaningful improvements in livelihoods for smallholder farmers in ways we couldn't have imagined twenty years ago.

In Nigeria, the *Cassava Disease Surveillance* platform uses AI to help farmers identify diseases through their mobile phones. A farmer takes a photo of a diseased leaf, and within seconds, the app diagnoses the problem and suggests treatment. The result? Higher yields and reduced crop losses.

The pattern is becoming hard to ignore: farmers don't reject innovation —they reject *inconvenient innovation*.

In Kenya, Tanzania, and Uganda, *Farmer Field Schools* have helped farmers significantly increase their incomes by doing two simple things well: learning by doing and getting timely information. The learning is practical—farmers observe, test, compare, and decide. And when mobile-based information systems support that process, the gains spread faster.

Interestingly, the biggest improvements often show up where the barriers used to be strongest: women farmers and farmers with limited literacy—because the method does not depend on long manuals... it depends on experience, demonstration, and peer support.

Now consider mechanisation—one of the most stubborn bottlenecks in African agriculture. The problem has never been that farmers don't understand the value of tractors. The problem is that tractors are expensive, hard to access, and often unavailable exactly when the rains arrive and time becomes the enemy.

That is why models like *Hello Tractor* work. They don't ask every farmer to buy a tractor. They simply connect the farmer who needs ploughing to the tractor owner who needs business—turning idle capacity into productive work. It is "Uber for farmers" in the sense that it matches supply with demand, but the deeper point is this: **it solves a real problem in a way that fits farmers' lives.**

That is the first lesson—technology succeeds when it fits the farmer, not when the farmer must fit the technology. A simple tool that helps a farmer act today is more valuable than a complex system that requires constant internet connectivity in places where the network disappears as soon as you leave town.

Precision Agriculture: Doing More With Less

The next frontier is not more inputs—it is better decisions. That is what precision agriculture really means: using data to make every seed, every bag of fertiliser, and every labour hour count.

In Ethiopia, Nigeria, and Mali, digital agronomy tools such as *RiceAdvice* have provided farmers with location-specific recommendations—when to plant, how much fertiliser to apply, and how to manage the crop based on local conditions. Across multiple settings, farmers have reported meaningful yield improvements in cereals like rice and wheat, alongside better profitability—not because they worked harder, but because they wasted less.

This is the difference between *prescription* and *precision*. Prescription says, "Apply this rate everywhere." Precision asks, "What does *your* soil need? What does *your* rainfall pattern allow? What does *your* variety respond to?"

In parts of West Africa and the Sahel, digital soil mapping initiatives are helping farmers see what used to be invisible soil constraints, nutrient gaps, and the exact kind of fertiliser response their land is likely to give. It reduces guesswork. It reduces waste. And it makes small budgets work harder and better.

That is the **second lesson**: **precision beats prescription.** Generic advice often wastes resources. Specific, data-driven guidance tends to protect farmer cash and maximise returns.

The Organic Premium: When Quality Pays

Remember those farmers at Brong-Densuso? They weren't just trying to control pests. They were trying to produce organic cocoa for premium prices in European markets.

Here's what they discovered: Organic production isn't just about avoiding chemicals. It's about understanding ecology. The farmers who learned to manage capsids using neem extract, pheromones, and natural predators didn't just reduce costs—they produced a higher-value product that commanded better prices.

This pattern is repeating across Africa. Farmers who can meet quality standards and certifications are accessing premium markets. The challenge isn't demand—European and Asian consumers are willing to pay more for sustainably produced African crops. The challenge is helping farmers meet those standards and connecting them to those markets.

The lesson: Quality creates value. But farmers need support to achieve quality standards and access premium markets.

The Cooperative Advantage: Strength in Numbers

In Ghana, a cocoa cooperative enrolled 50 farmers in an "Organic Cocoa Production" course delivered via mobile phones with offline capability and live Q&A workshops. Within six months:

- Yields increased by 30%
- Costs dropped by 20% (switching from chemicals to neem-based pesticides)
- The cooperative expanded exports to EU markets

Why did it work? Because cooperatives solve multiple problems at once:

- Economies of scale: Buying inputs together reduces costs
- Shared learning: Farmers learn from each other's successes and failures
- Market access: Cooperatives can negotiate better prices and access markets individual farmers can't reach
- Risk sharing: When one farmer's crop fails, the cooperative provides a safety net

The lesson: Individual farmers are vulnerable. Organised farmers have countervailing powers.

Youth and Agribusiness: the New Entrepreneurs

In Nigeria, 100+ rural youth launched agricultural micro-enterprises after completing an "Agribusiness Planning" course, everything from poultry to seed distribution. Average annual revenue per participant? $5,000.

This is the future of African agriculture: young people who see farming not as subsistence but as business. They're using technology, accessing finance, and building brands. They're not just growing food —they're building companies.

The lesson: When agriculture is profitable and modern, young people will choose it. When it's backbreaking and unprofitable, they'll leave for the cities.

What Makes These Solutions Work?

Looking across all these success stories, five common factors emerge:

1. Farmer-Centred Design
Solutions that work are designed *with* farmers, not *for* them. They solve real problems farmers face, not problems researchers think they should have.

2. Appropriate Technology
The best technology isn't the most advanced—it's the most appropriate. A simple mobile app that works offline beats a sophisticated system that requires constant connectivity.

3. Community Learning
Knowledge spreads fastest through trusted networks. Farmer-to-farmer learning beats top-down extension every time.

4. Economic Incentives
Farmers adopt practises that improve their bottom line. If it doesn't make economic sense, it won't scale - no matter how technically sound.

5. Institutional Support
Even the best farmer-led innovations need support: access to inputs, credit, markets, and enabling policies.

The Convergence Opportunity

Here's what excites me most: These solutions work even better when combined. Imagine a farmer who:

- Belongs to a cooperative (strength in numbers)
- Uses a mobile app for pest diagnosis (appropriate technology)
- Participates in a LARC (farmer-centred research)
- Produces for premium markets (economic incentive)

- Receives targeted support from extension services (institutional backing)

This isn't fantasy. It's happening in pockets across Africa. The question is: How do we scale it?

The Scaling Challenge

Success stories are inspiring. But they're also frustrating. Because for every farmer using AI to diagnose crop diseases, there are thousands still losing crops to pests they can't identify. For every cooperative accessing premium markets, there are hundreds of farmers selling at exploitative prices to middlemen.

The gap between what's possible and what's happening is enormous. And closing that gap requires more than just good technology or smart farmers. It requires a fundamental shift in how we think about agricultural development.

That paradigm shift is what we'll explore in the next chapter: moving from scattered successes to systemic transformation.

But first, remember this: The solutions exist. They're proven. They're working.

The question isn't, "Can Africa feed itself?"

But rather the question should be, "Will we scale what works?"

Key Takeaways:

- Participatory approaches like LARC dramatically improve knowledge transfer and adoption

- Mobile technology and AI are solving real problems for smallholder farmers

- Precision agriculture increases yields while reducing input costs

- Cooperatives provide economies of scale and market access

- Youth see agriculture as a business opportunity when it's profitable and modern

- Solutions work best when they're farmer-centred, economically viable, and institutionally supported

Chapter 4

Policy That Works for Farmers, Not Just Politicians

Let me tell you about the day I watched a government policy nearly destroy four years of work.

It was 2002-2005, and I was at Brong-Densuso, working with cocoa farmers who had spent years developing organic pest management practices. They'd formed cooperatives, learned new techniques, built relationships with international buyers. They were *this close* to certification—to finally getting the premium prices that would reward their hard work and environmental stewardship.

Then the government announced the "free" cocoa mass spraying programme.

Sounds good, doesn't it? Free pest control for farmers. What politician wouldn't love that headline?

But here's what the policy didn't account for: those synthetic pesticides would kill the predatory ants the farmers had been cultivating. It would contaminate their organic cocoa. It would destroy their certification prospects. Years of work, gone because the policy was designed in an air-conditioned office in Accra, not in consultation with the farmers it was supposed to help.

That's the difference between policy that looks good in a press release and policy that actually works in the field.

The Policy Gap

Here's the uncomfortable truth: most agricultural policies in Africa are designed to solve political problems, not farming problems.

Some politicians need to show they're "doing something" about food security. So, they import tractors without knowing who needs them, announce subsidies, mass spraying programmes, price controls— interventions that make for good photo opportunities, but these often make farmers' lives harder.

According to the World Bank, subsidies represent the most common form of government response to agricultural challenges in Africa, and input subsidies represent the largest share of those expenses. Yet, these same subsidies often crowd out more productive investments in research, extension, and infrastructure.

The result? In 2021, African governments spent an average of just 4.1% of their budgets on agriculture—less than half the 10% target they committed to under the Comprehensive Africa Agricultural Development Programme (CAADP).

What Farmers Actually Need from Policy

During my years working with farmers, from cocoa growers in Ghana to tomato farmers across West Africa, I've learned that effective agricultural policy has five essential characteristics:

1. It's Designed WITH Farmers, Not FOR Them

Remember those LARC farmers I mentioned? The Local Agricultural Research Committees we established didn't just implement research—they *designed* it. The community selected representatives, identified their most pressing problems, and co-created solutions with scientists.

The results were remarkable and sustained to date. When we surveyed farmers after two years, those who had participated in or been exposed to the LARC approach had significantly better knowledge, practices, and attitudes about pest management than those who had only received top-down extension messages.

Why? Because the research addressed *their* priorities, in *their* context, using *their* knowledge as a starting point.

Policy works the same way. The African Union's new CAADP Strategy 2026-2035 recognises this, emphasising inclusive design processes that ensure farmers are partners, not just beneficiaries.

2. It Addresses the Whole System, Not Just One Piece

Here's what I learned from that mass spraying debacle: you can't solve a pest problem without considering the market system, the certification requirements, the farmer organisations, and the ecological context.

Good policy recognises that agriculture is a *system*. You can't fix production without fixing markets. In fact, the market must precede production. You can't improve yields without addressing land tenure. You can't increase incomes without investing in processing and value addition.

Uganda's President Yoweri Museveni put it bluntly at the 2025 AU Summit: "The battle for value addition has been a big one because interest groups want to keep Africa as a raw-materials-producing continent. Adding value to agricultural products ensures vertical integration in the agricultural sector—from the garden to the table and from the farm to the wardrobe."

He's right. Policy that only focuses on production is policy that keeps farmers poor.

3. It's Flexible Enough To Adapt to Local Conditions

One of the biggest problems with the mass spraying programme was its rigidity. Same pesticides, same timing, same approach—whether you were growing organic cocoa at Brong-Densuso or conventional cocoa in Ashanti Region.

But agriculture isn't uniform. Soils differ. Climates vary. Markets change. Farmer capacities and aspirations are diverse.

Effective policy provides frameworks and resources, but allows for local adaptation. It recognises that what works in Kenya's highlands might not work in Mali's Sahel. It gives farmers and local officials the flexibility to respond to their specific conditions.

4. It Invests in What Actually Drives Productivity

Here's a sobering statistic: one dollar invested in agricultural research yields, on average, benefits equivalent to $10. Investments in irrigation can also yield high returns.

Yet most African governments continue to pour money into untargeted subsidies that provide temporary relief but don't build long-term productivity.

The World Bank report mentioned earlier is clear: countries need to reallocate public support from subsidies toward innovation systems, skills development, and productive infrastructure that enhance resilience and agricultural productivity.

That means:

- Research systems that work with farmers to develop appropriate technologies
- Extension services that facilitate learning, not just deliver messages
- Rural infrastructure—roads, storage, processing facilities—that reduce post-harvest losses and connect farmers to markets
- Climate-smart agriculture investments that boost yields while addressing soil and water degradation
- Digital infrastructure that enables farmers to access information, services, and markets

5. It Creates an Enabling Environment for Private Investment

Governments can't and shouldn't do everything. The private sector has a crucial role to play in agricultural transformation. But private investment requires an enabling environment: secure property rights, transparent regulations, responsive banking, functioning markets, reliable infrastructure, and policies that don't change with every election.

According to the Commercial Agriculture for Smallholders and Agribusiness (CASA) programme, Africa faces an agricultural financing gap of between $27 billion and $65 billion annually. Bridging that gap requires both public investment and policies that attract and support responsible private investment.

Effective policy also requires honesty about why most agricultural policies underperform — and the honest answer is often political economy, not technical design. A policy that would genuinely benefit smallholder farmers may threaten input dealers who profit from inefficient supply chains. A price reform that rewards farmers may reduce the leverage of buying companies with political connections. A transparent permitting system may close the space that informal networks currently occupy.

This is not to say reform is impossible — the cocoa producer price increase described later in this chapter demonstrates that transformative policy change can happen. But it happened partly because the political cost of inaction had become higher than the political cost of reform. Understanding that logic — who gains, who loses, and what shifts the calculus — is as important as understanding the technical content of the policy itself. Reformers who ignore political economy design brilliant policies that never survive contact with reality.

The Policy Reforms That Are Actually Working

Not all policy news is bad. Some African countries are getting it right.

Here's what's working:

Ghana's Producer Price Increases
When Ghana increased the cocoa producer price from 30% to 70% of the Free on Board (FOB) price between 2002 and 2004, production more than doubled. Farmers responded to better incentives—proof that getting prices right matters.

Rwanda's Agricultural Transformation
Rwanda is among the top African countries on track to achieve all the CAADP Malabo commitments by 2025. How? By consistently investing in agriculture, supporting farmer cooperatives, improving land tenure security, and building rural infrastructure. It's not magic—it's sustained political commitment and smart policy.

The African Continental Free Trade Area (AfCFTA)

The AfCFTA offers enormous potential to enhance food security through regional trade. Currently, only 15% of Africa's exports go to other African countries, and only 15% of that is agricultural products.

Imagine if we could triple intra-African agricultural trade, which is exactly what the new CAADP strategy aims to do by 2035. Farmers in surplus regions could supply deficit regions. Countries could specialise in what they grow best. Regional value chains could develop.

But it requires removing the non-tariff barriers, the arbitrary export bans, the border delays that currently make it easier to trade with Europe than with your neighbour.

Kenya's Digital Agriculture Initiatives

Kenya has led the way in using mobile technology to deliver agricultural services—from M-Pesa for payments to iCow for livestock management to M-Farm for market information.

These aren't just apps—they're policy choices. Kenya created a regulatory environment that enabled mobile money. It invested in digital infrastructure. It supported innovation.

The Institutional Challenge

Here's what I learned from my research: even good policies fail without strong institutions to implement them.

During our work at Brong-Densuso, we needed support from multiple institutions: the Cocoa Research Institute of Ghana (CRIG), the Ministry of Food and Agriculture (MoFA), the Ghana Cocoa Board (COCOBOD), the District Cooperative Office, and private sector partners.

Sometimes we got that support. Often, it depended more on individual champions within those organisations than on the systems themselves. When key people moved on, support evaporated.

That's not sustainable.

Effective agricultural policy requires:

- Coordinated action across ministries and agencies
- Adequate resources for implementation, not just policy formulation
- Skilled personnel at the national and local levels
- Accountability mechanisms that track results, not just spending
- Continuity across political transitions

The new CAADP strategy recognises this, calling for strengthened mutual accountability through biennial agricultural review processes and enhanced capacity for knowledge and data generation.

The Land Tenure Elephant in the Room

We can't talk about agricultural policy without addressing land tenure.

At Brong-Densuso, we discovered that one of the biggest barriers to improved pest management wasn't technical—it was social. The malfunctioning tenancy agreements between landlords and caretaker farmers meant that caretakers had no incentive to invest in long-term improvements like pruning or uprooting diseased trees.

Why would you invest in land you might lose next season?

This pattern repeats across Africa. Insecure land tenure—especially for women and youth—discourages investment, limits access to credit, and perpetuates low productivity.

Countries that have addressed land tenure—like Rwanda with its systematic land registration—have seen significant agricultural gains. Those that haven't continue to struggle.

Climate Policy and Agriculture

At the 2025 Africa Climate Summit in Addis Ababa, African leaders adopted the Africa Action Plan on Carbon Markets. It's a recognition that climate change is already affecting African agriculture—and that agriculture must be part of the climate solution.

But here's the challenge: climate policies often treat agriculture as a problem (a source of emissions) rather than as a solution (a potential carbon sink and adaptation strategy).

Farmer-centred climate policy would:

- Support climate-smart agriculture practices that increase productivity while building resilience
- Invest in climate information services that help farmers make better decisions
- Develop insurance and social protection systems that help farmers manage climate risks
- Ensure that carbon markets benefit smallholder farmers, not just large corporations
- Prioritise adaptation investments in the most vulnerable farming communities

From Policy to Practice: Making It Happen

So how do we move from policy documents to actual change in farmers' fields?

Based on my experience, here's what works:

1. Start with Pilot Projects
Test policies at small scale before rolling them out nationally. Learn what works, what doesn't, and why. Adapt based on evidence.

2. Build Coalitions
Effective policy reform requires champions across government, civil society, the private sector, and farmer organisations. Build those coalitions early.

3. Invest in Implementation Capacity
The best policy in the world fails without people and systems to implement it. Invest in training, systems, and resources.

4. Create Feedback Loops
Establish mechanisms for farmers to provide feedback on policy implementation. Adjust based on what you learn.

5. Measure What Matters
Track outcomes that matter to farmers—incomes, yields, resilience— not just outputs like number of trainings or tons of fertiliser distributed.

6. Be Patient but Persistent
Agricultural transformation takes time. But it requires sustained commitment, not stop-start initiatives that change with every election.

The Bottom Line

Good agricultural policy isn't complicated. It's policy that:

- Listens to farmers

- Addresses real constraints

- Invests in what works

- Creates enabling environments

- Builds strong institutions

- Adapts to local contexts

- Measures real outcomes

The question isn't whether we know what to do. We do. The question is whether we have the political will to do it—to design policy that works for farmers, not just for politicians.

Because here's the truth: when farmers succeed, everyone benefits. Food security improves. Rural economies grow. Youth find opportunities. The environment becomes more sustainable.

But when policy fails farmers, we all pay the price—in hunger, in poverty, in migration, in conflict.

The choice is ours.

Partnerships and policy influence

Illegal mining is not only local; it is systemic. STRED supports national and international partners working on coordination, enforcement systems, and restoration pathways.

- Policy coordination support
- Stakeholder alignment workshops
- Implementation roadmaps

www.stredconsult.com

Solutions scale through partnership >

Chapter 5

Bridging the Research Practice Divide

"The problem isn't that we don't have solutions. The problem is that our solutions are sitting in laboratories, books and reports while farmers are struggling in the field."

I will never forget the day an experienced cocoa farmer at Brong-Densuso taught me more about capsid pests than I'd learned in years of formal research.

We were standing in his farm, and he pointed to what he called the "cocoa mosquito"—an insect he'd been battling for years. According to our research at CRIG, this particular species caused minimal damage. But this farmer had built his entire pest management strategy around it.

That moment crystallised something I'd been sensing for years: there was a massive gap between what researchers knew and what farmers practised. And the tragedy was that both sides had valuable knowledge the other desperately needed.

The Expensive Silence Between Lab and Land

Africa spends billions on agricultural research every year. Our universities are filled with brilliant scientists. Our research institutes produce hundreds of studies annually. Yet, adoption rates for new agricultural technologies across the continent hover between 0.4% and 3.5%.

Let that sink in. For every 100 technologies we develop, fewer than four make it to farmers' fields.

This isn't just an academic problem. It's a crisis that costs us harvests, incomes, and food security. While researchers publish papers about drought-resistant varieties, farmers watch their crops wither. While we perfect organic pest control methods in controlled trials, smallholders spray expensive chemicals that poison their soil and their health.

The research-practice divide is one of the most expensive silences in African agriculture.

Why the Gap Exists

The divide didn't happen by accident. It's built into the way we've structured agricultural research for decades.

The Linear Trap

For too long, we've operated on what I call the "linear model": researchers develop technologies, extension agents deliver them, and farmers adopt them. It sounds logical. It's also fundamentally broken.

This model assumes that researchers are the sole custodians of knowledge, that extension is merely a delivery mechanism, and that farmers are passive recipients waiting to be enlightened. Reality is far more complex.

When I started working with cocoa farmers in the Eastern Region, I quickly learned that they weren't empty vessels waiting to be filled with our wisdom. They were sophisticated experimenters who had been managing complex agroecological systems for generations. They knew things about local conditions, seasonal patterns, and pest behaviour that no laboratory study could capture.

The problem wasn't that farmers lacked knowledge. Yes, though they may have gaps in the knowledge systems. The problem was that we weren't listening.

The Ivory Tower Problem

Research agendas are often drawn up in conference rooms far from the field. Priorities are set by funding agencies, international donors, and academic committees—rarely by the farmers who will ultimately use the technologies.

I've sat in countless meetings where we debated research priorities without a single farmer in the room. We made what some call "pre-analytic choices"—fundamental decisions about what problems to study and how to study them—without consulting the people most affected by those choices.

The result? We develop solutions to problems farmers don't have, while ignoring the challenges that keep them awake at night.

The Adoption Obsession

When technologies fail to spread, the question we typically ask is: "How do we make farmers adopt our recommendations?"

This is the wrong question. It assumes the problem lies with farmers rather than with the technologies themselves or the process that produced them.

The right question is: "Why aren't our technologies working for farmers?" This shifts the focus from farmer behaviour to research relevance.

What Actually Works: The Convergence Approach

After years of frustration with low adoption rates, I joined a research initiative that took a radically different approach. We called it the "Convergence of Sciences"—but it was really about the convergence of knowledge, stakeholders, and perspectives.

Starting With Farmers' Realities

Instead of beginning with what we wanted to research, we started with what farmers needed. We spent months at Brong-Densuso and surrounding communities, not lecturing but listening.

We used participatory tools—community meetings, seasonal calendars, problem-tree analyses—to understand farmers' priorities. We didn't just ask what problems they faced; we asked them to rank those problems, explain their causes, and propose potential solutions.

What emerged was fascinating. The constraints farmers identified weren't always what we expected. And their proposed solutions often drew on indigenous knowledge we'd never considered.

The Cage Experiment: When Farmers Become Scientists

Remember that "cocoa mosquito" I mentioned? The farmer's misconception about which insect caused the most damage wasn't just wrong—it was leading to ineffective pest management across the entire community.

But instead of simply telling farmers they were mistaken, we designed an experiment together. We collected different capsid species from their farms—the cocoa mosquito (*Helopeltis* spp.), the brown capsid (*Sahlbergella singularis*), and the black capsid (*Distantiella theobroma*). We set up cages containing cocoa pods and shoots, introduced various insects, and monitored damage over 96 hours.

Farmers participated in every step: collecting specimens, setting up the experiment, recording observations, and analysing results.

The outcome was transformative. When farmers saw with their own eyes that the cocoa mosquito caused minimal damage while the brown and black capsids were the real culprits, they didn't just change their minds—they changed their practices. And because they'd discovered this through their own experimentation, the knowledge stuck.

This is what I call "discovery learning"—and it's far more powerful than any extension pamphlet. Fast forward to 2006-2007, when I was hired by Trade and Investment Program for Competitive Export Economy (TIPCEE) sponsored by USAID, I used the same method to design a maize demonstration plot countrywide. A practice that was later replicated by ACDI-VOCA at scale in Ghana.

The LARC Model: Farmers as Research Partners

We took this participatory approach further by establishing Local Agricultural Research Committees (LARCs)—small groups of farmers elected by their communities to represent them in the research process.

These weren't token consultations. LARC members were genuine research partners. They helped design experiments, conducted field trials, analysed data, and reported findings back to their communities.

The results were remarkable. Knowledge and practices that typically remained confined to a small group of trained farmers spread rapidly through the community. Complex principles about pest ecology, which usually don't transfer well, were being discussed and applied by farmers who'd never attended a formal training

Why? Because the knowledge came from their peers, was tested under their conditions, and addressed their priorities. Science was democratised; no more monopoly by a few in research stations.

The Institutional Barriers We Must Overcome

Even when participatory research succeeds at the field level, it often crashes against institutional walls.

The Policy Problem

In the middle of our organic cocoa research, the Ghanaian government launched a nationwide mass spraying campaign using synthetic pesticides. Overnight, our farmers' prospects for organic certification—and their motivation to adopt integrated pest management—were threatened.

We had to mobilise all stakeholders to negotiate an exemption for our study area. It worked, but it highlighted a crucial point: good research isn't enough if policies undermine it.

Agricultural policies in many African countries still treat trees, crops, and livestock as separate domains. Forestry departments manage trees. Agriculture departments manage crops. Livestock departments manage animals. Nobody manages the integrated systems that farmers actually practise.

Until we break down these institutional silos, agroforestry and other integrated approaches will struggle to scale.

The Extension Gap

Extension services across Africa are chronically underfunded and understaffed. In Ghana's cocoa regions, extension agents are responsible for thousands of farmers spread across vast areas. They lack transportation, resources, and often the training to support complex innovations like integrated pest management.

Moreover, extension systems are still largely designed for one-way technology transfer. They're structured to deliver messages, not facilitate learning. They reward the number of farmers reached, not the quality of outcomes achieved.

We need to reimagine extension as a facilitation service—helping farmers experiment, learn, and adapt—rather than a delivery mechanism for predetermined packages.

The Research Incentive Problem

Academic researchers are rewarded for publishing papers, not for impact on farmers' lives. A scientist who spends years working with communities to develop locally adapted solutions may publish fewer papers than one who conducts controlled laboratory experiments. The living testimony is that the Cocoa Organic Farmers Association (*earlier known as TOFA)* at Brong-Densuso, where my PhD field experiments were carried out, is still practising organic cocoa farming and exporting for a premium. Not just that, the practice has scaled by expanding into nearby communities. That is impact.

Until research institutions value participatory engagement and real-world impact as much as they value publications, the research-practice divide will persist.

Lessons from the Field

After years of participatory research, several principles have become clear:

1. Farmers are experimenters, not adopters

Stop thinking about "technology adoption" and start thinking about "farmer innovation." Farmers don't passively adopt technologies; they actively experiment, adapt, and integrate new practices into their existing systems.

Our role as researchers isn't to develop finished products for farmers to adopt. It's to provide options, principles, and knowledge that farmers can experiment with and adapt to their specific conditions.

2. Context is everything

What works in one location may fail in another—not because the technology is flawed, but because contexts differ. Soil types, rainfall patterns, market access, labour availability, cultural practices, and land tenure arrangements all shape what's feasible and desirable.

This is why participatory research, grounded in local conditions, is essential. You can't design context-appropriate solutions from a distance.

3. Social innovation matters as much as technical innovation

Many agricultural challenges aren't primarily technical—they're social and institutional. Land tenure insecurity, gender inequalities, weak farmer organisations, and dysfunctional value chains often constrain productivity more than lack of technical knowledge.

In our cocoa research, we discovered that mistrust between landowners and tenant farmers was undermining pest management. No amount of technical training would solve that problem. We needed new social arrangements and institutional innovations.

4. Integration beats isolation

The most successful interventions integrate multiple elements: improved varieties, better management practices, farmer organisation, market linkages, and policy support.

Single-component solutions—a new seed variety, a pest control method, a fertiliser recommendation—rarely transform farming systems on their own. Farmers need integrated packages that address multiple constraints simultaneously.

5. Learning platforms accelerate impact

Farmer Field Schools, LARCs, and Community IPM platforms all share a common feature: they create spaces for collective learning and experimentation. These platforms don't just transfer knowledge—they build farmers' capacity to analyse problems, test solutions, and make informed decisions.

Investing in these learning platforms may be more cost-effective than traditional extension in the long run, because they create lasting capacity rather than temporary compliance.

Scaling What Works

The challenge now is scaling these participatory approaches beyond pilot projects.

Institutionalizing Participation

Research institutes need to move beyond seeing themselves as "expert institutes" and embrace their role as "learning institutes" where farmers and other stakeholders are genuine partners in knowledge creation.

This requires changes in research protocols, funding mechanisms, staff incentives, and organisational culture. It's not easy, but it's necessary.

Building Bridges

We need stronger linkages among research, extension, farmer organisations, private-sector actors, and policymakers. Quarterly inter-institutional meetings, joint field visits, and collaborative planning sessions can help build these connections.

In our cocoa work, bringing together researchers, extension agents, organic marketing companies, and farmer associations created synergies that none could achieve alone.

Investing in Facilitation Skills

Participatory research requires different skills than conventional research. Scientists need training in facilitation, participatory methods, systems thinking, and stakeholder engagement.

Similarly, extension agents need to shift from being messengers to being facilitators of farmer learning and experimentation.

Creating Enabling Policies

Governments need to create policy environments that support integrated, farmer-centred approaches. This means:

- Recognising and rewarding participatory research in academic evaluation systems
- Funding extension services adequately and restructuring them to support farmer learning
- Ensuring agricultural policies are coherent across sectors (agriculture, forestry, environment, trade)
- Involving farmers in policy design, not just implementation

The Path Forward

Bridging the research-practice divide isn't just about better communication or more extension agents. It requires fundamentally rethinking how we generate and share agricultural knowledge.

It means recognising that farmers are knowledge creators, not just knowledge users. It means designing research with farmers, not just for them. It means valuing impact on livelihoods as much as impact on academic journals.

Most importantly, it means humility. As researchers, we must acknowledge that we don't have all the answers. The solutions to Africa's agricultural challenges will emerge from the creative collaboration between scientific knowledge and farmer innovation.

The good news? When we get this right, the results are transformative. Technologies that work under farmers' conditions. Practices that farmers actually want to use. Knowledge that spreads organically through communities. Impact that lasts.

The research-practice divide isn't inevitable. It's a choice—a choice about how we organise research, who we include in the process, and what we value as success.

It's time to make a different choice.

Visit *bit.ly/landandlegacybooks2026* to get the companion book, Gold or Life? Illegal Mining and the Fight for Ghana's Land, Youth, and Legacy

Chapter 6

Sustainable Intensification - The Only Path Forward

The old man at Brong-Densuso had a simple way of explaining what was happening to his cocoa farm. "The soil is tired," he told me, gesturing at the sparse canopy above us. "We've been asking it to give and give, but we haven't been giving back."

He was right, of course. But his observation pointed to something bigger than just his farm or even his community. It captured the central dilemma facing African agriculture today: how do we produce more food for a growing population while healing—not harming—the land that sustains us?

The answer isn't to farm less. It's to farm smarter. And that's what sustainable intensification is all about.

The False Choice

For too long, we've been presented with a false choice: either we intensify agriculture to feed our people, or we protect the environment. Either we embrace modern inputs and technology, or we stick with traditional methods. Either we prioritise productivity, or we prioritise sustainability.

But these aren't either-or propositions. They are not mutually exclusive; they never were.

During my years working with cocoa farmers, I've seen what happens when we chase productivity at any cost. I've watched farmers spray synthetic pesticides four times a season because that's what they were told to do, only to see their beneficial insects disappear, their soil health decline, and their costs spiral upward. I've seen yields drop despite—or perhaps because of—increased chemical use.

I've also seen what happens when we romanticise traditional farming without acknowledging its limitations. Yields stagnate. Farmers remain poor. Young people flee to the cities. And ironically, desperate farmers end up clearing more forest to compensate for low productivity on existing land.

The path forward isn't about choosing between these extremes. It's about transcending them entirely.

What Sustainable Intensification Actually Means

Sustainable intensification means producing more from the same land while improving—not degrading—the natural resources that farming depends on. It means higher yields *and* healthier soils. More income *and* greater resilience. Increased productivity *and* enhanced ecosystem services.

It sounds almost too good to be true. But I've seen it work.

In our research at Brong-Densuso, we worked with farmers to test alternative methods for controlling capsids—the insects that can devastate cocoa yields. Instead of relying solely on synthetic pesticides, we explored three approaches: neem extract as a botanical pesticide, sex pheromone traps for population management, and encouraging populations of *Oecophylla longinoda*, a predatory ant that naturally controls capsids.

The results were remarkable. Not only did these methods effectively manage the pest, but they also cost less, posed no health risks to farmers, and maintained the ecological balance of the farm. Farmers who adopted these practices saw their yields stabilise or increase while their input costs decreased. They were producing more with less —and doing it sustainably.

That's sustainable intensification in action.

The Core Principles

Based on research across Africa and my own experience in Ghana, sustainable intensification rests on several core principles:

1. Work with nature, not against it

The most successful farmers I know are keen observers of natural systems. They understand that healthy soil is alive—teeming with microorganisms that cycle nutrients, improve structure, and suppress diseases. They recognise that biodiversity isn't just nice to have; it's essential for pest control, pollination, and resilience.

Sustainable intensification harnesses these natural processes rather than trying to override them with external inputs. It means using cover crops to fix nitrogen instead of relying solely on synthetic fertilisers. It means encouraging beneficial insects instead of killing everything with broad-spectrum pesticides. It means building organic matter in the soil instead of mining it year after year.

2. Integrate crops, trees, and livestock

The most productive and resilient farms aren't monocultures. They're diverse systems where different components support each other.

In many parts of Africa, agroforestry—integrating trees with crops—has proven transformative. Trees provide shade for crops like cocoa and coffee, fix nitrogen, prevent erosion, diversify income, and sequester carbon. In Ethiopia, farmers practising agroforestry have reduced soil erosion by up to 50% while increasing crop yields.

Similarly, integrating livestock with crop production creates synergies. Animals provide manure for soil fertility. Crop residues feed animals. The system becomes more than the sum of its parts.

3. Optimise inputs, don't maximise them

More isn't always better. The goal isn't to use as much fertiliser or pesticide as possible—it's to use the right amount, at the right time, in the right place.

This is where precision matters. Soil testing tells you what nutrients your soil actually needs, so you're not wasting money on unnecessary inputs. Scouting for pests helps you determine whether intervention is even needed, and if so, what the most targeted approach would be.

In our capsid research, we found that farmers who learned to scout for pests and understand their life cycles could time interventions much more effectively. They used less, but achieved better results.

4. Close nutrient cycles

One of the biggest challenges in African agriculture is nutrient depletion. We harvest crops and remove nutrients from the farm, but we don't replace them adequately. Over time, soils become exhausted.

Sustainable intensification means closing these nutrient cycles. It means returning crop residues to the soil. Composting organic waste. Using legumes to fix nitrogen. Applying manure strategically. Minimising nutrient losses through erosion and leaching.

When farmers in Western Kenya adopted integrated soil fertility management—combining organic and inorganic inputs strategically— maize yields increased by an average of 70%. The soil got healthier, not poorer.

5. Build resilience, not just productivity

Climate change means that farming is becoming more unpredictable. Droughts, floods, pests, and diseases are all becoming more variable and severe.

Sustainable intensification builds resilience into farming systems. Diverse cropping systems are less vulnerable to any single shock. Healthy soils with high organic matter hold more water during droughts and drain better during floods. Integrated pest management reduces the risk of pest outbreaks.

Resilience also means economic diversification. Farmers who grow multiple crops, keep some livestock, and perhaps process some of their produce, aren't devastated when one enterprise fails.

The Technologies That Make It Possible

Sustainable intensification isn't about going backwards to some imagined agricultural past. It's about moving forward with the best of traditional knowledge *and* modern science.

Improved Varieties

New crop varieties—developed through both conventional breeding and biotechnology—offer higher yields, better nutrition, and greater resilience to drought, pests, and diseases. In Senegal, drought-tolerant millet varieties have helped farmers maintain production despite increasingly erratic rainfall.

The key is ensuring these varieties are developed *with* farmers, not just *for* them, so they meet real needs and fit into existing farming systems.

Conservation Agriculture

Minimum tillage, permanent soil cover, and crop rotation—the three pillars of conservation agriculture—reduce erosion, improve soil health, and save labour. In Zambia, farmers practising conservation agriculture have seen maize yields increase by 20-50% while reducing their labour requirements.

Precision Agriculture

Satellite imagery, soil sensors, and mobile apps are bringing precision to smallholder farming. Farmers can now access weather forecasts, market prices, and agronomic advice on their phones. They can use satellite data to monitor crop health and target interventions.

These technologies democratize information that was once available only to large commercial farmers.

Biological Inputs

Biofertilisers, biopesticides, and other biological inputs offer alternatives to synthetic chemicals. They're often cheaper, safer, and more environmentally friendly. The neem extract we tested in our cocoa research is one example. Rhizobium inoculants that enhance nitrogen fixation in legumes are another.

Water Management

Simple technologies like drip irrigation, rainwater harvesting, and water-efficient crop varieties can dramatically improve productivity in water-scarce environments. In Burkina Faso, farmers using planting pits (*zai*) to harvest rainwater have rehabilitated 300,000 hectares of degraded land and increased food production by 80,000 tons annually.

The Economics Make Sense

One of the most common objections to sustainable intensification is that it's too expensive or complicated for smallholder farmers. But the economics actually favour these approaches.

Yes, there may be upfront costs—for training, for new seeds, for soil amendments. But the returns are substantial and sustained.

Farmers practising sustainable intensification typically see:

- **Higher yields** that are more stable over time
- **Lower input costs** as they reduce dependence on expensive external inputs
- **Premium prices** for products certified as organic, fair trade, or sustainably produced
- **Reduced risk** from more resilient, diversified systems
- **Improved soil health** that compounds benefits over time

In our work with organic cocoa farmers, those who successfully adopted sustainable practices earned premium prices that were 20-30% higher than conventional cocoa. Even without the premium, their lower input costs meant better profit margins.

The Challenges Are Real

I won't pretend sustainable intensification is easy. The challenges are significant:

Knowledge gaps
These approaches require more knowledge and management skill than conventional farming. Farmers need to understand soil biology, pest ecology, and system interactions. Extension services need to shift from delivering prescriptions to facilitating learning.

Labour requirements
Some sustainable practices—like composting, mulching, or managing cover crops—can be labour-intensive, at least initially. This is a real constraint, especially for women farmers who already face time poverty.

Transition period
When farmers shift from conventional to sustainable practices, there's often a transition period where yields may dip before they recover. Farmers need support to get through this period.

Market access
Sustainable intensification works best when farmers can access markets that reward quality and sustainability. Without these market linkages, the economic incentives are weaker.

Policy environment
Government policies often favour conventional intensification—subsidising synthetic fertilisers and pesticides, promoting

monocultures, and neglecting research on sustainable alternatives. This creates an uneven playing field.

Why It's the Only Path Forward

Despite these challenges, sustainable intensification isn't just one option among many. Given the realities of climate change, land scarcity, environmental degradation, and rural poverty, it's really the only viable path forward. Consider the alternatives:

Conventional intensification—based on high external inputs is becoming economically unviable for smallholders as input prices rise. It's also environmentally unsustainable, degrading the very resources farming depends on. And it's increasingly risky as climate change makes farming more unpredictable.

Extensification—expanding the area under cultivation—isn't an option in most of Africa. We've already converted most of the suitable land. Further expansion means destroying forests, wetlands, and other critical ecosystems. It's neither environmentally nor economically sustainable.

Low-input traditional farming—may be sustainable in the narrow sense of not degrading resources, but it doesn't produce enough to lift farmers out of poverty or feed growing populations. It's not economically or socially sustainable.

Sustainable intensification is the only approach that addresses all dimensions of sustainability—environmental, economic, and social. It's the only approach that can feed Africa's growing population while healing degraded lands, adapting to climate change, and providing decent livelihoods for farmers.

What Success Looks Like

I've seen glimpses of what's possible when sustainable intensification is done right.

I've seen cocoa farmers at Brong-Densuso producing organic cocoa that earns premium prices while maintaining healthy, biodiverse farms.

I've seen maize farmers in Kenya doubling their yields through integrated soil fertility management while improving their soil health.

I've seen vegetable farmers in peri-urban areas using drip irrigation and organic inputs to produce high-value crops year-round.

I've seen pastoral communities integrating improved forages and better livestock management to increase productivity while reducing pressure on rangelands.

These aren't isolated success stories. They're proof of concept. They show what's possible when we combine the best of traditional knowledge with modern science, when we work with nature rather than against it, and when we support farmers as innovators rather than treating them as passive recipients of technology.

The question isn't whether sustainable intensification can work in Africa. It's whether we have the political will, the institutional capacity, and the financial resources to scale it up.

That's what the next chapters are about.

Key Takeaways:

- Sustainable intensification means producing more while improving natural resources

- It's based on working with nature, integrating systems, optimising inputs, closing nutrient cycles, and building resilience

- Modern technologies—from improved varieties to precision agriculture—make it possible

- The economics favour sustainable approaches over the long term

- Despite real challenges, it's the only viable path forward for African agriculture

The Digital Revolution on African Farms

The phone rang at 6:47 a.m. It was Akua, a cocoa farmer from the Ashanti Region.

"Dr Ayenor, I just checked the app. It says capsid pressure is high this week. Should I spray the neem extract now or wait until after it rains?"

I smiled. Five years ago, Akua couldn't have asked that question. Not because she didn't know about capsids—she'd been battling them for twenty years. But because she had no way to access real-time pest forecasts, no platform to ask questions, and no confidence that anyone would answer.

Now, through her basic smartphone and the training system her cooperative purchased from programme's similar to STRED's Digital Farming and Development Institute, Akua has what extension officers used to provide to only a privileged few: timely information, expert advice, and a community of fellow farmers facing the same challenges.

This is the digital revolution on African farms. And it's not coming—it's already here.

The Mobile Moment We Almost Missed

Hopefully, what I am about to share here will excite every agricultural development professional: Africa has over 500 million smartphone users, and that number is growing by 20 million every year.

Think about what that means. In villages where there's no extension officer, no agricultural college, no library—there are smartphones. In communities where the nearest paved road is 30 kilometres away—there's mobile network coverage. In farms where farmers have never received a single training, there are young people scrolling through TikTok and WhatsApp.

We spent decades trying to build physical extension systems that could reach every farmer. We tried but collectively failed. The ratio of extension officers to farmers in most African countries is 1:1,000 or worse. In some areas, it's 1:5,000. Even the best extension officer can't effectively serve 5,000 farmers scattered across hundreds of square kilometres.

But mobile technology changes *everything*.

When I worked with FAO in 2018 on an institutional analysis of Ghana's e-agriculture system, and later in 2022 as a consultant for the Digital Villages Initiative (DVI), we assessed African countries' readiness to leverage digital tools for agricultural transformation. What we found was both exciting and frustrating.

The infrastructure is there. The farmers are ready. The technology works. But the content, the platforms, and the business models? Those are still catching up.

That's when we realised something critical: *the future of agricultural training isn't exactly about building another online university. It's about building the infrastructure that governments, NGOs, and agribusinesses can use to train farmers at scale.*

STRED's Digital Farming and Development Institute (SDFI): Enterprise Training Infrastructure for Africa

Let me be honest with you. When we started planning SDFI in 2023, people thought we were crazy.

"Farmers and rural population don't want to learn from phones," they said. "They need hands-on training."

"Rural internet is too slow," others warned. "Your platform will never work."

"Who will pay for it?" investors asked. "Farmers can't afford subscriptions."

They were asking the wrong questions.

The right questions were: *Who is already trying to train farmers, especially the young ones, at scale? What challenges do they face? And how can we build a system that solves those challenges?*

The answer became clear... Governments running agricultural programmes. Donor-funded projects. NGOs implementing value chain interventions. Agribusinesses managing outgrower schemes. Large cooperatives are strengthening their members.

They all need the same thing: a way to efficiently deliver quality training to thousands of farmers, track learning outcomes, and measure adoption—without building everything from scratch.

That's what SDFI became. Not an online school selling courses to individual farmers. *A training infrastructure system that organisations purchase as a vehicle or means to deliver their own programmes.*

How SDFI Enterprise Works: B2B2F Training at Scale

Here's the model that's transforming agricultural training across Africa:

Step 1: An Organisation Identifies a Training Need

Maybe it's a government ministry rolling out a new agricultural policy and needing to train 50,000 farmers across 10 regions. Maybe it's an agribusiness company with 5,000 outgrowers who need training in Good Agricultural Practices. Maybe it's a donor-funded project that needs to deliver climate-smart agriculture training with measurable outcomes.

Step 2: They Partner with STRED

Instead of building their own learning management system, hiring content developers, and figuring out how to deliver training on low-bandwidth networks—they license SDFI Enterprise.

Step 3: We Configure the System for Their Needs

We work with them to:
- Customise learning tracks to their specific crops, regions, and objectives
- Localise content into relevant languages
- Brand the platform with their identity

- Set up dashboards to track learner progress and outcomes
- Train their field officers to facilitate learning

Step 4: Training Reaches Farmers at Scale

Farmers receive training through:
- Mobile-first courses that work offline and on basic smartphones
- WhatsApp-based learning communities for peer support
- Field facilitators who guide hands-on practice
- AI-powered chatbots that answer questions 24/7 in local languages

Step 5: Impact is Measured and Reported

The organisation gets:
- Real-time dashboards showing enrollment, completion, and engagement
- Adoption tracking (are farmers actually applying what they learned?)
- Impact reports for donors and stakeholders
- Case studies and testimonials for communications

Three Ways Organisations Use SDFI Enterprise

Package 1: SDFI Enterprise – Cohort Delivery (Pilot)

Best for: First-time buyers testing digital training

A cocoa cooperative in Ghana wants to train 200 members in organic production methods. They purchase a 6-month pilot with STRED.

What they get:

- 3 customised learning tracks (pest management, soil health, post-harvest handling)
- Learner onboarding and completion tracking
- Simple assessments and certificates
- WhatsApp community facilitation
- End-of-cohort impact report

Result: 30% yield increase, 20% cost reduction, expanded exports to EU markets.

Package 2: SDFI Enterprise – Programme Rollout (Scale)

Best for: Government programmes and multi-district projects

The Ministry of Agriculture is implementing a national youth agribusiness programme across 50 districts. They need to train 10,000 young farmers in entrepreneurship, production, and market access.

What they get:

- 8 learning tracks tailored to different value chains
- Training-of-trainers for district agricultural officers
- Field adoption support (demo plots, peer champions)
- Monthly dashboards and MEL integration
- Quarterly learning reviews

Result: 100+ youth-led agribusinesses launched, average revenue $5,000 per participant annually.

Package 3: SDFI Enterprise – Transformation Partnership (12 Months)

Best for: Large agribusinesses and long-term government programmes

A multinational food company has 15,000 smallholder farmers in its supply chain across three countries. They need continuous training, quality improvement, and traceability.

What they get:

- Full learning calendar with continuous onboarding
- Role-based pathways (farmers, lead farmers, cooperative leaders, field officers)
- Custom modules branded as "[Company Name] Farmer Academy"
- Advanced analytics linking training to yield and quality outcomes
- Annual impact report with case studies and communications assets

Result: 25% yield increase, improved quality standards, reduced supply chain losses, enhanced brand reputation.

Why This Model Works: The B2B2F Advantage

Traditional agricultural training fails for three reasons:

1. It's expensive to scale – Sending trainers to every village costs too much
2. It's hard to track – Did farmers actually learn? Are they applying it? Requires a mid-term evaluation which also adds to the budget

3. It's not sustainable – When donor funding ends, the training stops - we see this over and over again in our evaluations

SDFI Enterprise solves all three:

Scalable: Once content is developed, it can reach 1,000 farmers or 100,000 farmers at nearly the same cost.

Measurable: Every interaction is tracked. Organisations know exactly who completed what, when, and whether they're applying it.

Sustainable: Organisations pay for the system as part of their programme budgets. No dependency on external donors for STRED's operations.

The Technology Behind the Training

People often ask: "How does this work in rural areas with poor internet?" Here's how we designed SDFI to work in African realities:

Mobile-First Design

- Works on the most basic smartphones
- Videos compressed to use minimal data
- Content can be downloaded and viewed offline
- Zero-rated access through telecom partnerships (farmers don't use their data)

AI-Powered Support

- Chatbots answer common questions instantly in local languages
- Image recognition helps farmers diagnose pest and disease problems
- Predictive analytics alerts farmers to upcoming risks

Blended Learning

- Not just phone-based: combines mobile content with field facilitators
- Demo plots where farmers practice what they learned
- Peer learning groups for community support

Data Protection

- Farmers own their data
- Transparent consent processes
- Secure, encrypted systems
- Compliance with international data protection standards

The 2026 Vision: Training 100,000 Farmers Through Institutional Partnerships

Here's where we're headed:

Target for 2026:

- 3-6 enterprise contracts signed (mix of pilot and scale packages)
- 10,000+ farmers trained through institutional partners
- 1 anchor client on a 12-month transformation partnership
- 2 public case studies demonstrating measurable impact

By 2027:

- 100,000 farmers reached across 10 African countries
- SDFI is recognised as the leading enterprise training infrastructure for African agriculture

- Partnerships with major donors, governments, and agribusinesses

By 2030:

- 1 million farmers trained
- SDFI becomes the standard for digital agricultural extension across Africa

What This Means for African Agriculture

The digital revolution isn't about replacing extension officers or eliminating in-person training. It's about augmenting human capacity with technology so that:

- One extension officer can effectively support 5,000 farmers instead of 500
- A cooperative can train all its members simultaneously instead of in small batches
- A government programme can reach every district with consistent, quality training
- An agribusiness can ensure all outgrowers meet quality standards
- Farmers get the right information at the right time, in their language, on their phone

This is the convergence of:

- Mobile technology (the infrastructure)
- Quality content (the knowledge)
- Institutional partnerships (the scale)
- Measurable outcomes (the accountability)

The Path Forward: Five Priorities for Scaling Digital Agriculture

Based on my experience with FAO's Digital Villages Initiative, Ghana's e-agriculture strategy, and SDFI's growth, I see five priorities:

1. Invest in Digital Infrastructure—Governments and telecom companies must expand network coverage, increase speeds, and reduce data costs in rural areas. This is foundational.

2. Develop Local Content—We need thousands of videos, courses, and tools in local languages, addressing local crops and local challenges. This requires investment in content creation and partnerships with agricultural research institutions.

3. Build Digital Literacy—Schools, extension services, and community organisations must teach farmers how to use smartphones and digital services effectively. This is as important as teaching agronomic practices.

4. Create Enabling Policies—Governments must develop regulations that protect farmer data, promote interoperability between platforms, and incentivise private sector investment in digital agriculture.

5. Develop Sustainable Business Models—Digital agriculture services must generate revenue to be sustainable. The B2B2F model—where institutions pay for training infrastructure—is proving more sustainable than trying to charge individual farmers.

A Vision of the Future

Let me end with a vision of what's possible. Imagine a government ministry that needs to train 50,000 farmers in climate-smart agriculture. Instead of spending years building a system from scratch, they partner with STRED and launch the programme in 90 days.

Imagine an agribusiness company that can trace every farmer in its supply chain, verify their training status, and link training completion to quality improvements and premium payments.

Imagine a young woman in rural Ghana who has never attended university. Through her smartphone and her cooperative's SDFI-powered training programme, she completes courses in sustainable cocoa production, farm business management, and digital marketing. She applies what she learns, doubles yields, and triples income.

Through her farmers' group or a firm and the SDFI, she uses the platform to connect directly with chocolate buyers in Europe who pay premium prices for her sustainably produced cocoa. She accesses credit through a fintech app to invest in processing equipment. She starts producing cocoa butter and chocolate bars, creating jobs for other young people in her community.

She becomes a mentor on the platform, sharing her knowledge with thousands of other farmers. She's invited to speak at agricultural conferences. She starts her own agribusiness consulting firm.

This is the future we're building. And it could be closer than you think.

The digital revolution on African farms is about more than technology. It's about democratizing knowledge, empowering farmers, and unleashing the potential that's been locked away for too long.

The infrastructure is ready. The model is proven. The partners are waiting. Now let's scale it.

Fintech and Agricultural Finance Innovation

One of the biggest barriers to agricultural transformation in Africa: access to finance.

Banks don't usually want to lend to smallholder farmers. The loans are too small to be profitable. The risks are too high. Farmers lack collateral and credit histories. And monitoring thousands of small loans across rural areas is expensive and difficult.

So, farmers can't access the capital they need to buy quality inputs, invest in equipment, or expand their operations. They're trapped in a cycle of low productivity and poverty.

Digital technology is breaking this cycle.

During my work with the International Finance Corporation from 2018 to 2020, I was involved in Ghana's Warehouse Receipt System Project. The concept is simple but powerful: farmers store their crops in certified warehouses and receive receipts that can be used as collateral for loans.

But making it work required digital systems to track inventory, verify quality, connect farmers with lenders, and facilitate transactions. We developed training programmemes to help farmers understand the system. Basically, how a commodity structured trading system works. We worked with banks to design appropriate loan products. And demonstrated how they could use mobile money to disburse loans and collect repayments.

COVID-19 somewhat disrupted the full success story. But the initial results were impressive. Farmers who participated in this type of financing arrangement under the warehouse receipt system accessed credit at interest rates lower than traditional sources (which was still high). They could store their crops and sell when prices were high, rather than being forced to sell immediately after harvest at low prices. And default rates were below 5%, demonstrating that smallholder farmers are creditworthy when the system is properly designed.

Similarly, under TechnoServe Ghana, I was privileged to manage what was then known as the Inventory Credit Program for maize between 1997 and 2001 in Ghana's maize belt.

But warehouse receipts are just one innovation. The real revolution is in digital credit scoring and mobile lending.

Traditional banks assess creditworthiness based on collateral, income statements, and credit history—things most smallholder farmers don't have. But farmers do have other data: mobile phone usage patterns, input purchase history, land ownership records, social networks, and farming practices.

Fintech companies are using this alternative data to create credit scores for farmers who have never had a bank account. And they're disbursing loans directly to farmers' mobile money accounts, with repayments deducted automatically at harvest time.

At SDFI, we intend to partner with fintech providers to integrate credit access into our platform. A farmer completes a course on improved maize production. The system assesses their knowledge and commitment. A lender offers them a small loan to purchase certified seeds and fertiliser. The farmer uses the inputs, increases their yield, and repays the loan from the additional income. Their credit score improves, and they qualify for a larger loan next season.

This is financial inclusion at scale. And it's only possible because of digital technology.

DIGITAL INNOVATION SPOTLIGHT

Scan the QR code below to:

- Watch a demo of SDFI Enterprise in action
- Download the "Digital Training Readiness Assessment" for your organisation
- View case studies from our pilot partners
- Request a consultation for your agricultural programme

Visit: https://stredconsult.com/about-sdfi/

Building the Ecosystem - What Governments Must Do

During one of our field visits to the Western Region, although we were in a Toyota Fortuna, we noticed the roads were poor. A farmer we met was desperate and frustrated with the poor condition of the road. "Dr Ayenor, they promised us the roads would be fixed three years ago. My dried cocoa beans are absorbing moisture and getting rotten because I can't get them to the Licensed buyer market. What good is a 70% producer price if I can't sell my beans?"

This conversation haunts me because it captures a fundamental truth about agricultural transformation: the government's role is not to do everything but to create the enabling environment in which farmers, businesses, and communities can thrive and operate efficiently.

Too often, we see governments trying to be the farmer, the trader, the processor, and the regulator all at once. They fail at all of them. The most powerful thing a government can do is build the ecosystem—the infrastructure, policies, and frameworks that allow everyone else to succeed.

Let me be clear, from my many years working with governments across Africa, Latin America and Europe: the best agricultural policies are those that get out of farmers' way while paving the way ahead of them.

The Infrastructure Imperative: Roads, Storage, and Processing

In my 2023 Cabinet paper on agricultural financing, the evidence showed that post-harvest losses in Ghana are commonly estimated at 30–40% across many perishable value chains—driven largely by gaps in storage, cold-chain, transport, and market infrastructure, rather than by on-farm production alone.

Think about that. A farmer can do everything right—use improved seeds, practice integrated pest management, harvest at the right time —and still lose nearly half the value of their crop because:

- The road to the market is impassable during the rainy season
- There's no storage facility within 50 kilometres
- The nearest processing plant is in another region entirely

The Road Less Travelled (Because It Doesn't Exist)

During our work on the Marshall Plan for Agricultural Transformation report in 2018, I shared this experience on field assessments conducted earlier across Ghana's agricultural zones. In the Northern Region, I met a women's cooperative growing tomatoes. Their yields had tripled using improved varieties and drip irrigation.

But every rainy season, they watched 50% of their harvest rot because the 15-kilometre road to the nearest market became a river of mud.

"We don't need more training," the cooperative leader told me. "We need a road."

She was right. Infrastructure is not just about convenience—it's about whether agricultural transformation is even possible.

The infrastructure priorities that emerged from my policy work with multiple African governments are clear:

1. Farm-to-Market Roads—Not only highways. Not only urban expressways. All-weather rural roads that connect production zones to collection points and markets are equally important. In Ghana's cocoa sector, when we improved feeder roads in the Eastern Region, farmers' net incomes from selling cocoa and other farm produce increased by 23% within one year—not because yields improved, but because post-harvest losses dropped and farmers could negotiate better prices by reaching markets quickly. In Ghana, improvements to feeder roads in the regions increased farmers' net incomes within a year, not through higher yields, but by reducing post-harvest losses and enabling quicker access to markets and better prices.

2. Strategic Storage Facilities—Under the Ghana Warehouse Receipt System Project that I worked on with the International Finance Corporation, we discovered that farmers with access to certified warehouses earned more for the same crops. Why? Because they could store their harvest and sell when prices were favourable, rather than dumping everything on the market at harvest time when prices crash.

The warehouse receipt system (or Inventory Credit Program, as earlier referred to during Technoserve) does something even more powerful: it turns crops into collateral. A farmer with a warehouse receipt can access credit using their stored maize as security. This is financial inclusion through infrastructure.

3. Processing Facilities at Scale—Africa's exports are mostly raw materials, and that accounts for about 65 % of its agricultural commodities. We ship cocoa beans and import chocolate. We export cashew nuts and import cashew products. Every time we do this, we export jobs, value, and economic transformation.

The government's role is not to build and operate processing plants—that's where the private sector excels. The government's role is to:

- Provide reliable electricity and water to industrial zones

- Streamline licensing and permits for agro-processing businesses

- Invest in shared infrastructure (effluent treatment, cold chains) that individual businesses can't afford alone

Governments need to create the conditions for private processors to thrive.

Regulatory Reforms: Removing the Barriers We Created

Many of the biggest obstacles to agricultural transformation are policies we created ourselves.

Land Tenure: the Silent Killer of Investment

In my PhD research on cocoa production, I documented how land tenure uncertainty devastates long-term agricultural investment. A farmer who doesn't have secure rights to their land will not:

- Plant tree crops that take 3-5 years to mature

- Invest in soil improvement that pays off over decades

- Build anything permanent; certainly not any infrastructure

In one community I studied, 28% of cocoa farmers were "caretakers" managing land for absentee landlords. Their share of the harvest? Typically one-third. Their incentive to invest in improving productivity? Zero.

Securing land tenure and proper documentation are as important as the price increase.

The policy recommendations made to multiple governments are consistent:

- Digitise land records to reduce disputes and corruption

- Recognise customary land rights rather than trying to replace them with Western models

- Protect tenant farmers with minimum tenure periods that make investment worthwhile

- Streamline land registration so it takes weeks, not years

Trade Policies: Opening Doors, Not Building Walls

The ECOWAS seed and fertiliser harmonisation work I contributed to through the MIRA Project revealed how our own trade barriers hurt farmers. I learnt how a Ghanaian farmer near the Burkina Faso border couldn't legally buy superior tomatoes or sorghum seeds from 20 kilometres away across the border, but could import inferior seeds from Europe.

Regional integration through AfCFTA creates unprecedented opportunities, but only if we reform the policies that make cross-border agricultural trade a nightmare.

In an ILO evaluation covering its 2018–23 programmes in West Africa, we heard how domestic trade can be slowed by numerous checkpoints and informal fees. One tomato trader described stopping at roughly a dozen-plus checkpoints on the Northern Ghana–Accra route and paying multiple 'fees' along the way—costs that can wipe out margins on perishable produce.

Smart trade policy reforms include:

- Harmonising standards across regions so farmers can access larger markets

- Eliminating internal trade barriers that make domestic trade harder than international trade

- Strategic tariff protection for infant industries while avoiding the trap of permanent protectionism

- Phytosanitary systems that protect against pests without becoming tools for corruption

Standards That Enable, Not Exclude

In my work on organic cocoa certification, I saw how standards can either open markets or lock smallholders out. European food safety standards are necessary, but when compliance requires equipment costing more than a smallholder's annual income, those standards become barriers.

The government's role is to invest in shared compliance infrastructure:

- Testing laboratories accessible to smallholder cooperatives

- Training programmes on good agricultural practices

- Certification support for farmer organisations

- Traceability systems that work for small-scale producers

Investment Frameworks: Making Public-Private Partnerships Actually Work

Most public-private partnerships (PPPs) in agriculture across Africa fail. The few that succeed share common characteristics that governments must understand.

The Ghana Warehouse Receipt System: a PPP Success Story

As a consultant for the IFC's Ghana Warehouse Receipt System Project between 2018 and 2020, I've seen how well-designed PPPs can transform agricultural finance. The model is elegant:

- Government provides the regulatory framework and warehouse certification

- Private sector builds and operates warehouses

- Financial institutions lend against warehouse receipts

- Farmers get fair prices and access to credit

By 2021, over 15,000 farmers were expected to have used the system, accessing millions in credit. The government's investment? Primarily, regulatory infrastructure and supporting the project's capacity building —a fraction of what it would cost to run a state-owned warehouse system.

The PPP Principles That Work

From my contribution as a Consultant to a team, and later as collaborator working with the $30M+ agribusiness financing mechanisms for USAID - Mobilising Finance for Agriculture, the World Bank and others, these principles emerge:

1. Clear Risk Allocation—Each partner takes the risks they're best positioned to manage. Government handles policy risk. Private sector handles commercial risk. Farmers handle production risk. When everyone tries to offload their risk onto others, the partnership collapses.

2. Genuine Additionality—PPPs should do things that neither public nor private sector can do alone. If the private sector would do it anyway, why is government involved? If government could do it better alone, why complicate it with private partners?

3. Farmer-Centric Design—I've seen too many PPPs designed around what's convenient for government and business, with farmers as an afterthought. The best PPPs start with the question: What do farmers need? Then design backward from there.

4. Long-Term Commitment—Agricultural transformation doesn't happen in 2-year project cycles. The PPPs that work have 10-15 year horizons with patient capital and realistic expectations.

Case Study: How Policy Change Doubled Cocoa Production in Two Years

Let me take you inside one of the most dramatic agricultural policy successes in recent African history—one I had a front-row seat to observe and document in my research.

The Context: a Sector in Crisis

By the early 2000s, Ghana's cocoa sector was under severe strain. Output had fallen sharply from earlier levels, many farmers were losing confidence in cocoa, and smuggling across borders was a growing concern. For many of us, it felt like an industry that helped build Ghana was slipping.

The problem was simple: In the early 2000s, Ghana's system was designed to pass a majority of the FoB price to farmers—often well over half—yet when world prices fell, and costs rose, many farmers still felt it wasn't enough to justify investment.

The Policy Intervention: a Radical Shift

In 2003, under pressure from the World Bank and facing an economic crisis, the Ghanaian government made a bold decision: it adhered to its promise to implement the policy to increase the producer price to 70% of FOB over two years.

This wasn't just a price adjustment. It was a fundamental rethinking of government's role. For decades, COCOBOD had operated on the principle that government knew best how to manage cocoa revenues. The price increase acknowledged a different truth: farmers are the best investors in cocoa production when they have the incentive to invest.

The Results: Beyond Expectations

The numbers tell a stunning story. A published source reports the following results for the cocoa crop seasons:

- 2001/02: 340,600 MT
- 2002/03: 496,846 MT (+46%)
- 2003/04: 736,911 MT (+48%)

In two years, production more than doubled. But here's what's even more remarkable: This happened *before* most of the government's other interventions (mass spraying, fertiliser subsidies, etc.) had taken full effect. The price increase alone unlocked massive latent production capacity.

What Actually Happened on the Ground

In my field research during this period, I documented what farmers did with better prices:

Immediate Actions (2003):
- Harvested cocoa that they had previously left to rot because transport costs exceeded the value
- Cleared overgrown farms that had been abandoned
- Applied basic maintenance (weeding, pruning) that they'd skipped when prices were low

Medium-Term Investments (2004-2005):
- Purchased improved seedlings to replace old, unproductive trees
- Invested in pest management (even before the mass spraying programme)
- Hired labour for intensive farm maintenance
- Expanded cultivation into previously marginal areas

The Lesson: Incentives Over Everything

This case study demolishes several myths about African agriculture.

Myth 1: "Farmers don't respond to price signals". Reality: Farmers are exquisitely sensitive to prices. They're running businesses, not charities. When the price is right, they invest. When it's not, they rationally allocate their labour elsewhere.

Myth 2: "Production increases require massive government programmes". Reality: The biggest increase in production came from simply paying farmers fairly. The expensive mass-spraying programme achieved full coverage of the cocoa belt by the time production had already doubled.

Myth 3: "Smallholders can't increase productivity quickly". Reality: When incentives align, smallholders can mobilise resources and knowledge remarkably fast. The 46% increase in one year shows how much "latent capacity" exists when farmers have been suppressed by bad policy.

The Caveat: Price Alone Isn't Enough

The cocoa price increase worked because other elements of the ecosystem were already in place:

- COCOBOD's quality control system ensured farmers could sell their increased production
- Decades of research by CRIG meant that improved varieties were available
- Basic infrastructure (feeder roads, buying stations) existed, even if inadequate
- Farmer's knowledge of cocoa production was high

Price was the catalyst, but it worked because the ecosystem was already in place to support the response. This is why I emphasise that the government's role is building the *ecosystem*, not just pulling one lever. The cocoa success came from getting the incentive structure right within a functioning (if imperfect) system.

The Government Scorecard: What to Measure, How to Hold Leaders Accountable

After authoring the agricultural and rural transformation chapters of Ghana's Coordinated Programme of Economic and Social Development Policies (2017-2024 and the revised 2021-2025 version), I learned that policies without accountability mechanisms are just expensive wishes.

Here's the scorecard I recommend every African government adopt for agricultural transformation:

Infrastructure Metrics (Measure What Matters)

DON'T MEASURE ⊖	DO MEASURE ☑
Kilometres of roads built	Percentage of farmers who can reach a market within 2 hours year-round
Number of warehouses constructed	Percentage of harvest that can be stored under certified conditions
Megawatts of electricity generated	Percentage of agro-processing zones with reliable 24/7 power

Regulatory Efficiency Metrics

DON'T MEASURE ⊖	DO MEASURE ☑
Number of new policies passed	Time and cost to register land, get permits, clear customs
Number of standards adopted	Percentage of smallholder cooperatives that can afford compliance

Investment Climate Metrics

DON'T MEASURE ⊖	DO MEASURE ☑
Amount of government spending on agriculture	Ratio of private investment to public investment in agriculture
Number of PPPs signed	Number of PPPs still functioning after 5 years

Outcome Metrics (The Bottom Line)

DON'T MEASURE ⊖	DO MEASURE ☑
Tons of fertiliser distributed	Yield increases per hectare
Number of farmers trained	Farmer incomes and poverty rates
Value of agricultural exports	Percentage of exports that are processed (not raw commodities)

The Accountability Mechanism

Metrics mean nothing without consequences. In my policy advisory work, I've advocated for:

1. Public Dashboards—Real-time data on key metrics, accessible to citizens. When farmers can see that their district has received 40% of promised infrastructure investment while another district got 120%, they ask questions.

2. Independent Evaluation—My work as an independent evaluator for organisations like CARE International sponsored by United States Department of Labour (USDOL), the ILO, and the European Commission has shown the power of external assessment. Governments should mandate independent evaluation of all major agricultural programmes and the ensuing report be used to improve the situation.

3. Farmer Feedback Loops—The best accountability comes from beneficiaries. Regular farmer surveys, grievance mechanisms, and participatory monitoring ensure that policies are judged by those they're meant to serve.

4. Political Consequences—Ultimately, accountability requires that leaders who fail to deliver face consequences not only at the ballot box, but also be prosecuted as necessary. This requires an informed electorate with access to credible data—which brings us back to public dashboards and independent evaluation.

The Enabling Environment: A Checklist for Leaders

Let me close this chapter with a practical checklist drawn from my experience working with governments beyond Ghana. If you're a policymaker, minister, or president, ask yourself:

Infrastructure:

- Can 80% of farmers reach a market within 2 hours year-round?
- Is there certified storage capacity for at least 30% of national production?
- Do agro-processing zones have reliable electricity and water?

Land Tenure:

- Can a farmer register land in less than 30 days?
- Are customary land rights legally recognised?
- Do tenant farmers have security for long-term investment?

Trade Policy:

- Can farmers access regional markets without prohibitive barriers?
- Are standards harmonised with major trading partners?
- Is the time to clear customs in less than 48 hours?

Investment Climate:

- Is private agricultural investment growing faster than public investment?
- Can a farmer cooperative access credit by using harvest as collateral?
- Are PPPs designed with clear risk allocation and farmer-centric goals?

Accountability:

- Are key metrics publicly available and updated regularly?
- Are major programmes (i.e., cocoa marketing 2015 -2025) independently evaluated?
- Do farmers have effective mechanisms to provide feedback and lodge complaints?

If you can't check most of these boxes, you're not creating an enabling environment—you're creating obstacles with good intentions.

The Government We Need

That farmer we met who was talking about the broken road? Six months later, after the government fixed the road, his cooperative's income increased significantly by the next season.

He didn't need the government to farm for him. He didn't need another training programme. He needed the government to do what only the government can do: build the road, set fair rules, and get out of the way. The role of the agriculture minister does not include transporting plantains to sell at the Ministry.

That's the government African agriculture needs. A government that creates an ecosystem where farmers, businesses, and communities can do what they do best.

Reflection Questions:

1. What infrastructure gaps in your region prevent farmers from capturing the full value of their harvest?

2. Which government policies in your country help farmers, and which ones hinder them? How can you advocate for reform?

3. If you're in government: Are you measuring what matters, or what's easy to measure?

4. How can citizens hold their governments accountable for promises of agricultural transformation?

Chapter 9

The Private Sector's Role - Beyond CSR

"The best development programme is a profitable business that treats farmers as partners, not charity cases."

I begin by sharing with you a meeting that changed how I think about agricultural development.

It was 2018, and I was sitting in a conference room with some key actors including representative of a Bank in Ghana, where I served as an Agricultural Development Consultant. Across the table sat a young entrepreneur who wanted a loan to start a warehouse receipt system for maize farmers in the Northern Region.

"Why should we fund this?" the bank staff asked. "There are NGOs doing farmer support. Why does this need to be a business?"

The entrepreneur's answer was brilliant: "Because NGOs leave. Donors get tired. Projects end. But a profitable business? That stays. That scales. That transforms."

He was right. And he got the loan.

That warehouse receipt system is now serving over 5,000 farmers, has mobilised millions in agricultural finance, and is expanding to three more regions. Not because of charity. Because it's profitable.

The CSR Trap

For too long, some, including banks, have treated private sector involvement in agriculture as Corporate Social Responsibility (CSR)—something companies do to look good, to tick a box, to satisfy stakeholders. A photo opportunity with smiling farmers. A donation of seeds or fertilisers. A training programme that lasts six months and disappears.

Don't get me wrong. CSR has its place. But it's not transformation.

Real transformation happens when the private sector sees agriculture not as a charity case but as a business opportunity. When companies invest not because it makes them look good, but because it makes them money. When farmers are partners in profitable value chains, not beneficiaries of handouts.

I've learned one fundamental truth: sustainable agricultural transformation is driven by profitable business models, not by good intentions.

The $1 Trillion Opportunity

Africa's food and agriculture market is projected to reach $1 trillion by 2030. One trillion dollars. That's not aid money. That's not donor funding. That's real market demand from real consumers willing to pay for food.

Right now, we're importing $78 billion worth of food annually. That's $78 billion leaving our economies, creating jobs in Europe, Asia, and the Americas instead of in Accra, Lagos, or Nairobi.

But here's the opportunity: every dollar we currently spend on imports is a dollar that could go to African farmers, African processors, African distributors, and African retailers.

The private sector doesn't need to be convinced there's a market. The market is already there. What's missing is the infrastructure, the systems, and the business investment models to capture it.

What Works: Profitable Models for Agricultural Transformation

Let me show you what works when the private sector gets it right.

1. Value Chain Integration: From Farm to Fork

In 2021, I was part of a team that profiled 67 agribusinesses across Ghana for USAID's Mobilising Finance for Agriculture project. These weren't just any businesses—they were companies that had figured out how to connect themselves and other smallholder farmers to profitable markets.

During the study, one company stood out: a cashew processor in the Bono Region, where TechnoServe gave me the opportunity to support the growth of cashew as a viable commodity in Ghana. Let me acknowledge the immense contributions of colleagues - Mr Tom Bonney, Mr John Addaquaye, Dr Emmanuel Foe, and others.

Instead of just buying raw cashews from farmers at whatever price middlemen offered, this company did something different:

- Backward integration: They provided farmers with improved seedlings, training, and access to credit
- Quality assurance: They established collection centres with moisture meters and grading systems
- Forward integration: They processed the cashews in Ghana, adding value before export
- Market linkages: They secured contracts with European buyers willing to pay premium prices for traceable, quality cashews

The result? Farmers' incomes increased by 25% - 40%. The company's profit margins improved. Ghana captured more value from its cashews. Everybody won.

This is what value chain integration looks like.

2. Warehouse Receipt Systems: Turning Crops Into Collateral

Between 2018 and 2020, I worked as a consultant for the IFC World Bank Group on the Ghana Warehouse Receipt System Project (*sponsored by IFC and SECO*). My role as a consultant-trainer ended with the COVID-19. This project addressed one of the biggest problems facing African farmers: they're forced to sell their harvest immediately after harvest when prices are lowest, because they have no storage and need cash urgently.

Here's how warehouse receipt systems change the game:

A farmer harvests 10 tons of maize. Instead of selling immediately at rock-bottom prices, he stores all or some in a certified warehouse. The warehouse issues him a receipt—a legal document proving he owns that maize.

That receipt becomes collateral. He can take it to a bank and get a loan for 70% of the maize's value. He uses that loan to pay school fees, buy inputs for the next season, or invest in his farm.

Three months later, when maize prices have increased by 30-40%, he sells the maize, repays the loan, and pockets the difference.

The warehouse makes money from storage fees. The bank makes money from interest. The farmer makes more money from better prices. The system works because everyone makes a profit.

We trained about 2,000 beneficiaries, including farmers, bankers, regulators, and 50 warehouse operators across Ghana. The results? Farmers who used the system increased their income by an estimated average of 35%. Banks that were previously reluctant to lend to farmers suddenly saw agriculture as a viable lending opportunity.

This is what happens when you create business models that align incentives.

3. Outgrower Schemes: Guaranteed Markets, Guaranteed Supply

Similarly, during my time with USAID's TIPCEE project (2006-2007), as the Lead on Food Crops, I led my team to design and establish demonstration plots connected to outgrower schemes for maize and tomato. From the demonstration fields, we linked them with buyers. The model is simple but powerful:

A processing company or exporter contracts with smallholder farmers to grow specific crops to specific standards. The company provides:

- Inputs (seeds, fertilisers) on credit

- Technical support through extension services provided by the USAID-TIPCEE field demonstration plots via discovery learning approaches

- Guaranteed market at pre-agreed prices

The farmers provide:

- Land and labour
- Commitment to follow quality standards
- Delivery of the agreed quantity

When done right, out-grower schemes solve multiple problems simultaneously:

- Farmers get access to inputs they couldn't afford upfront
- Farmers get technical support to improve quality and yields
- Farmers get guaranteed markets, eliminating price risk
- Companies get reliable supply of quality produce
- Companies can trace their supply chain for certification

Within Ghana's maize belt, while working for the United Nations World Food Programme (WFP), I linked over 4,000 maize farmers to various processing companies. They included Yedent and General Mills (now Premium Foods), major processors, who were in turn listed as food suppliers to WFP.

Farmers' incomes increased because they were selling directly to the processor without middlemen, and the school feeding and other humanitarian programmes benefited. There were a few contractual challenges between farmers and some companies, but they were surmountable.

That notwithstanding, here's the key: this only works when companies treat farmers as partners, not as exploitable labour. When contracts are fair. When payments are timely. When technical support is genuine, and farmers respond to the market without "side-selling".

4. Agribusiness as Development Partner

One of the most exciting shifts I've seen is companies recognising that investing in smallholder productivity isn't just good PR—it's good business.

Take the cocoa sector operating within semi-structured trading system. Companies like Kuapa Kokoo in Ghana have shown that when you invest in farmer cooperatives, provide fair prices, and share profits, you don't just help farmers—you secure a reliable supply of quality cocoa.

Despite its challenges, Kuapa Kokoo is a farmer-owned cooperative that operates as a Licensed Buying Company (LBC). It purchases cocoa from over 100,000 smallholder farmers and exports directly. But it goes further:

- Farmers own a significant share of The Day Chocolate Company (makers of Divine Chocolate)

- Social premiums fund community infrastructure—schools, water systems, health clinics

- Fair trade certification ensures farmers get better prices

- Democratic governance means farmers have a voice in decisions (this area is where many fail – accountability)

This is a business model that recognises farmers as stakeholders, not just suppliers. And it's profitable—for the farmers and for the company.

5. Digital Platforms: Connecting Farmers to Markets

The digital revolution is creating entirely new business models.

In Kenya, Twiga Foods uses mobile ordering and a modern logistics network to link smallholder suppliers with urban vendors/retailers. Farmers benefit from more reliable market access and faster digital payments, while retailers get more consistent supply and quality. Twiga earns revenue through service/margin and delivery-related fees tied to these transactions.

In Nigeria, Farmcrowdy has been reported to connect 'farm sponsors'—often urban investors—with farming cycles, channelling finance and inputs to farms and sharing returns at harvest. In this model, Farmcrowdy takes a share (e.g., 20%) as its platform revenue, while investors and farmers share the remainder. Farmers get capital they couldn't access from banks. Investors earn returns that are better than those of savings accounts. The platform earns fees.

As STRED's former client during its inception, AgroCenta LLC, with its technology platform, plays a similar role in Ghana as an active aggregator–sourcing farm produce–to improve agricultural commodity value chains in the country. These platforms work because they solve real problems—market access, finance, information—in ways that create value for everyone involved.

The Investment Opportunities: Where the Money Is

If you're an investor, an entrepreneur, or a business leader wondering where the opportunities are, let me be specific.

Based on my work profiling agribusinesses and conducting value chain assessments across West Africa, here are the highest-potential investment areas:

1. Post-Harvest Infrastructure

In Sub-Saharan Africa, post-harvest grain losses alone have been valued at about $4 billion a year—a major opportunity for storage, logistics, and processing solutions. That's not a problem—it's a $4 billion opportunity.

Investments needed:

- Storage facilities: Warehouses, cold storage, grain silos
- Processing equipment: Mills, dryers, packaging machines
- Transportation: Refrigerated trucks, logistics systems (cold chain)

The returns are substantial because you're capturing value that's currently being wasted.

2. Value Addition and Processing

We export raw cocoa beans and import chocolate. We export most of our raw cashews and import roasted cashews. We export raw cotton and import clothes.

Every step of processing we do locally is value we capture, jobs we create, and profits we keep.

Investments needed:

- Processing facilities: Cocoa processing, cashew processing, fruit processing, textile mills
- Quality control systems: Testing labs, certification support
- Branding and marketing: Building African brands that compete globally

3. Agricultural Finance

The African Development Bank estimates an annual agricultural financing shortfall of up to $65 billion in Africa... a major market opportunity for financial institutions willing to innovate.

Opportunities:

- Warehouse receipt financing: Lending against stored commodities
- Value chain financing: Financing entire supply chains, not just individual farmers
- Digital lending: Using mobile money and alternative credit scoring
- Agricultural insurance: Protecting farmers against weather and price risks

During my time at Development Bank Ghana, we identified scores of bankable agribusinesses that couldn't access finance simply because either banks didn't understand agriculture or they found them too risky. The opportunity is enormous for institutions willing to learn.

4. Input Supply and Distribution

Farmers need seeds, fertilisers, tools, and equipment. But rural distribution networks are weak, and prices are high.

Opportunities:

- Agro-dealer networks: Franchised input shops in rural areas
- Digital ordering systems: Farmers order inputs via mobile, delivered to their farms
- Equipment leasing: Tractors, irrigation systems, and processing equipment available for rent

5. Market Linkages and Aggregation

The biggest challenge for smallholders isn't production—it's market access. Businesses that solve this problem can be highly profitable.

Opportunities:

- Aggregation centers: Collecting produce from multiple farmers to achieve scale (i.e., AgroCenta)

- Digital marketplaces: Connecting farmers directly to buyers

- Export facilitation: Helping farmers meet international standards and access export markets

What Government Must Do to Enable Private Investment

Private sector investment doesn't happen in a vacuum. It requires an enabling environment. Given my extensive field experiences, including advising Ghana's Cabinet on agricultural financing and authoring national agricultural policies, here's what governments must do:

1. Infrastructure Investment

No business can thrive without basic infrastructure:

- Roads: Connecting farms to markets

- Electricity: Powering processing facilities and cold storage

- Water: Irrigation systems and processing needs

- Digital infrastructure: Internet connectivity in rural areas

These are public goods that governments must provide. The private sector can't do it alone.

2. Regulatory Reforms

Businesses need clear, stable, predictable rules:

- Land tenure security: So farmers and businesses can invest with confidence
- Trade policies: Reducing barriers to regional trade
- Standards and certification: Clear quality standards that are enforced fairly
- Contract enforcement: Legal systems that protect agreements

3. Risk-Sharing Mechanisms

Agriculture is risky. Government can help manage those risks:

- Agricultural insurance: Subsidising premiums to make insurance affordable
- Price stabilisation: Strategic reserves to smooth price volatility
- Credit guarantees: Sharing risk with banks to encourage agricultural lending

4. Public-Private Partnerships

The most successful agricultural investments combine public and private resources:

- Outgrower schemes: Government provides extension support, private sector provides markets
- Research and development: Public research institutions work with private companies such as STRED Consult.
- Infrastructure: Government builds roads, private sector builds warehouses

The Accountability Question

Here's the uncomfortable truth: not all private sector involvement in agriculture is positive. I've seen companies that:

- Exploit farmers with unfair contracts
- Grab land without adequate compensation
- Extract resources without investing in communities
- Make promises they don't keep

This is why we need accountability mechanisms:

- Transparent contracts: Farmers must understand what they're agreeing to
- Fair pricing: Mechanisms to ensure farmers get fair value
- Environmental standards: Companies must not degrade the resources they depend on (i.e., avoid water pollution)
- Social responsibility: Genuine investment in communities, not just photo opportunities

The private sector's role isn't to replace government or NGOs. It's to complement them. The government creates the enabling environment. NGOs build capacity and advocate for farmers. The private sector creates the profitable businesses that sustain transformation.

The Bottom Line

Agricultural transformation in Africa won't be driven by aid. It won't be driven by government programmes alone. It will be driven by profitable businesses that see farmers as partners, that create value for everyone in the chain, and that build systems that last.

The opportunity is enormous. The market is there. The farmers are ready.

The question is: Is the private sector, including development finance institutions and the banks, ready to move beyond CSR and embrace agriculture as the business opportunity of the century?

Because when we get this right—when we build profitable, inclusive, sustainable agricultural businesses—everyone wins. Farmers earn decent incomes. Companies make profits. Consumers get quality food. Economies grow. Communities thrive. Countries develop.

That's capitalism with social conscience at its best.

And that's the future we can build and sustain—if we have the courage to see agriculture not as a problem to be solved, but as an opportunity to be seized.

REFLECTION QUESTIONS

1. What are the most profitable agricultural value chains in your country? Who's capturing the value?

2. What infrastructure gaps are preventing private investment in agriculture in your region?

3. How can your organisation (business, government, NGO) create win-win partnerships with farmers?

4. What policy reforms would most encourage responsible private investment in agriculture?

5. If you're an entrepreneur, what agricultural problem could you solve profitably? If you're a farmer: What business partnerships would help you most?

STRED CONSULT CONNECTION

"Need help designing profitable, farmer-centred business models? STRED Consult specialises in value chain development, agribusiness profiling, and investment facilitation. Contact us to explore how we can support your agricultural investment strategy."

Visit: *https://stredconsult.com/agricultural-operations-and-value-chains/*

The New African Farmer - Youth, Women, and Innovation

The average age of African farmers is rising. In Ghana, it hovers around 55 years. In Kenya, it's 60. Across the continent, young people are fleeing rural areas for cities, viewing agriculture as their grandparents' profession—backbreaking, unprofitable, and decidedly unglamorous.

Meanwhile, women are central to Africa's food production and post-harvest activities, yet they make up only about 15% of agricultural landholders and receive less than 10% of agricultural loans in many smallholder farmers' contexts.

This is Africa's agricultural paradox: the people who could transform farming—youth with their digital fluency and entrepreneurial energy, women with their proven productivity, value addition and resilience—are systematically excluded from it.

The future of African agriculture doesn't just depend on better seeds or smarter policies. It depends on making farming attractive to the next generation and equitable for half the population. It depends on reimagining who an African farmer is.

The Youth Exodus: Why Young People Are Leaving

I've sat in too many village meetings where the youngest person in the room was over 40. The pattern is consistent across Africa: young people leave, and farming communities age.

The reasons are obvious when you listen to young people themselves. "Why would I farm?" a 24-year-old university graduate asked me in Kumasi, "My father works from sunrise to sunset and can barely pay my siblings' school fees. I'd rather drive for Uber."

He's not wrong to think this way. Traditional smallholder farming, as currently practised, offers:

- Low and unpredictable incomes
- Limited access to land (often controlled by elders in some areas)
- Backbreaking physical labour
- Social stigma ("farming is for those who couldn't do anything else")
- No clear path to prosperity

Yet agriculture remains Africa's largest employer and greatest economic opportunity. The African Development Bank estimates that agriculture and agribusiness could be a $1 trillion industry by 2030. The question isn't whether there's opportunity in farming—it's whether we can make that opportunity visible and accessible to young people.

Making Farming Cool: The Agripreneur Revolution

During my work with the Food and Agriculture Organization of the United Nations in 2022, I encountered over 100 agribusinesses seeking investment. What struck me wasn't just the diversity of enterprises—from shea butter, cassava and maize processing. It was the use of technology, including GPS, in locating smallholder farmers and the establishment of payment platforms. Elsewhere, you have emerging drone-based spraying services, but the entrepreneurs are young. Over 60% were under 35 years of age. These weren't traditional farmers. They were *agripreneurs*—young people who saw agriculture not as subsistence farming but as a business opportunity.

They were:

The Tech Integrators: Like Adjoa, 28, who started a mobile app connecting smallholder vegetable farmers directly to restaurants in Accra and door-to-door delivery of fresh vegetables, cutting out three layers of middlemen and tripling farmer incomes.

The Value-Add Innovators: across West Africa, young entrepreneurs are processing cassava into high-quality flour for bakeries, creating a shelf-stable product that can command higher prices than raw roots and support steadier demand.

The Service Providers: Like the trotro tractor service providers who run a tractor hire service in Northern Ghana, serving many smallholder farmers who can't afford their own equipment.

~

What these young entrepreneurs understood is that you don't have to be a farmer to be in agriculture. The value chain offers countless entry points: input supply, mechanisation services, processing, logistics, marketing, financial services, and technology solutions.

This is the reframing of agricultural needs. We're not asking young people to pick up hoes and replicate their grandparents' farming methods. We're inviting them to disrupt, innovate, and build profitable businesses in agriculture.

It is very similar to my son Jesse's innovative idea of building a platform, which he calls 'Acelink', to connect services of trusted handymen and women to homes, where real-time solutions are delivered instantly (i.e., plumbing).

The Business Case for Youth in Agriculture

In 2021-2022, I assisted Dr Mei Zegers in evaluating the European Union's Archipelago project in Ghana, which focused mainly on training programmes for youth entrepreneurship. Under the various TVET initiatives, I learned that young people don't need to be convinced that agriculture is important—they need to be shown that it's profitable. Looking into the future from then, I concluded the following and it is truly compelling:

Growing Markets: Africa's population will double by 2050. Food demand will triple. Someone will supply that food—why not young African entrepreneurs?

Technology Advantage: Young people are digital natives. They adapt quickly to precision agriculture tools, mobile money, e-commerce platforms, and data analytics—all of which are transforming modern farming.

Innovation Mindset: Unlike older farmers who may be set in traditional methods, young people question everything. "Why do we plant this way?" "Could we market differently?" This questioning leads to innovation.

What Young People Need to Enter Agriculture

Based on my work across multiple youth-focused agricultural projects, five factors consistently determine whether young people enter and stay in agriculture:

1. Access to Land—Traditional land tenure systems favour elders. Young people need alternative pathways: leasing arrangements, cooperative farming models, or government land allocation schemes specifically for youth.

2. Access to Finance—Banks view young people as risky borrowers. We need youth-specific financing products: smaller loans, longer grace periods, group lending models, and equity investment rather than just debt.

The reason I am excited that KPMG approached me to be part of the team selected to undertake a feasibility study on the establishment of Youth Entrepreneurship Investment Bank, sponsored by AfDB.

3. Skills and Knowledge—Not traditional farming knowledge, but business skills: financial management, marketing, negotiation, and digital literacy. The STRED Digital Farming and Development Institute was designed precisely for this—delivering modern agricultural business education via mobile phones.

4. Mentorship and Networks—Young agripreneurs need connections to markets, suppliers, and experienced business people. Incubation programmes and mentorship networks are critical.

5. Success Stories—Nothing attracts young people like seeing their peers succeed. We need to support and celebrate young agricultural entrepreneurs as loudly as we celebrate tech startup founders.

Women in Agriculture: The Productivity Multiplier

If youth are agriculture's future, women are its present—and its most underutilised asset.

As the lead evaluator for USDOL during my evaluation of CARE International's Adwuma Pa project, which focused on reducing child labour and improving conditions for women in Ghana's cocoa sector, I noticed something remarkable. When women received the same training, inputs, and support as men, their productivity sometimes matched or exceeded men's. Yet they continued to face systematic barriers. The statistics are stark:

- Women perform 60-80% of agricultural processing activities in Africa
- They manage household food security
- They reinvest over 70 -80%% of their income in family welfare (compared to 30-40% for men)
- Yet they own less than 15% of agricultural land
- They receive less than 10% of agricultural credit
- They have limited access to extension services, which typically target male "heads of household"

Gender Equity as Productivity Strategy

The Food and Agriculture Organisation estimates that if women farmers had the same access as men to productive resources—land, quality inputs, credit and extension—yields on their farms could rise by 20% to 30%. This would help to reduce the number of hungry people by roughly 100 million to 150 million.

I've seen this play out repeatedly. Under a USAID-sponsored TechnoServe's Cassava Processing activity (i.e., gari making at Akrofuom) undertaken by Catholic Women Association near Techiman and its environs, we deliberately targeted women-led enterprises for financing and technical support. Now, it is one of the biggest agro-processing activities in the area. The results:

- Higher loan repayment rates than those of male borrowers
- Faster business growth
- Greater community impact (because women's businesses tended to employ other women)
- More sustainable practices (women were more likely to adopt climate-smart techniques)

The barriers women face aren't mysterious:

Land Rights: In many African communities, women can't own land. They farm their husband's or father's land, which limits their ability to make long-term investments or use land as collateral for loans.

Time Poverty: Women carry the double burden of farm work and household responsibilities. They have less time for training, market visits, or cooperative meetings.

Social Norms: Some extension agents may ignore women farmers, assuming men make agricultural decisions. Some financial institutions view women as risky borrowers. Input suppliers usually target men in marketing products. Though this seems to be gradually changing, the process is still slow.

Limited Mobility: In some contexts, women face restrictions on travelling to markets or attending training sessions.

Addressing these barriers requires targeted interventions.

What Works: Lessons from the Field

Through my work on gender-focused agricultural projects, including evaluations of child protection and women's economic empowerment initiatives, I've seen what actually works:

1. Deliberate Targeting—Don't just say, "women can participate." Set specific targets: 50% of training participants must be women, 40% of loans must go to women-led enterprises. Measure and report on these targets.

2. Address Time Constraints—Hold training sessions at times and locations convenient for women or use online platforms like the SDFI for women. Provide childcare. Keep sessions short and practical.

3. Women-Only Spaces—In mixed groups, women often defer to men. Women-only farmer groups, savings groups, and cooperatives create space for women's leadership and voice.

4. Couple and Household Approaches—Gender dynamics play out at the household level. Engaging husbands and families in discussions about women's economic empowerment reduces resistance and increases support. However, these interventions must consider existing positive socio-cultural norms.

5. Link to Markets—Women often get stuck in low-value activities. Deliberately connecting women farmers to higher-value markets and value chains increases their incomes and status.

6. Legal and Policy Reform—Ultimately, sustainable change requires reforming discriminatory land laws, inheritance practices, and financial regulations.

The New African Farmer

I am inviting you to use your imagination and let me paint a picture of what the new African farmer looks like:

She's 28 years old. She has a secondary school education and a smartphone. She grows vegetables on two hectares of leased land, using drip irrigation and improved seeds. She learned production techniques from mobile-based courses and YouTube videos. She sells directly to restaurants via WhatsApp and receives payment through mobile money. She employs three other young women from her village. She's saving to buy a motorcycle for deliveries and eventually purchase her own land.

He's 32 years old. He doesn't own a farm—he owns a tractor and a sprayer. He provides mechanisation services to 150 smallholder farmers in his district, who book his services via a mobile app. He learned equipment maintenance through online tutorials. He's expanding into drone-based spraying services. He employs two young mechanics and is training them to eventually run their own service businesses.

They're 26 and 29, a couple who inherited a cocoa farm. Instead of just selling beans, they've started producing chocolate bars with distinctive local flavours—ginger, hibiscus, baobab. They sell online and at urban farmers' markets. They've become social media influencers, sharing their "farm to chocolate" journey with 50,000 followers. They're making ten times what their parents made from raw cocoa.

These aren't fantasies. These are real emerging patterns I've observed across Africa. They represent agriculture's future—if we create the conditions for them to thrive. Before we think of scale, these may look small shifts, but they are some of the strategic examples of how the new African can feed himself and expand.

What Needs to Happen

Making agriculture attractive to youth and equitable for women requires action at multiple levels.

For Governments:

- Reform land laws or adapt them to enable youth access and women's ownership

- Create youth-specific agricultural financing programmes

- Integrate agricultural entrepreneurship into school curricula

- Invest in rural infrastructure (roads, electricity, internet, schools, sports and entertainment centres) that makes rural life viable

- Set and enforce gender targets in agricultural programmes

For Development Partners:

- Fund youth and women-focused agricultural initiatives

- Support agricultural incubation and mentorship programmes

- Invest in digital agricultural education platforms

- Promote success stories of young agripreneurs

For Private Sector:

- Develop youth-friendly financial products

- Create market linkages for young and women farmers

- Invest in agribusiness startups

- Provide internships and training opportunities

For Communities:

- Challenge norms that exclude women from land ownership and decision-making
- Celebrate young people who choose agriculture
- Support women's farmer groups and cooperatives
- Create mentorship connections between generations

For Young People and Women:

- See agriculture as business opportunity, not just farming
- Invest in learning—formal and informal
- Form cooperatives and networks for collective strength
- Don't wait for perfect conditions—start small and learn as you go

The Demographic Dividend

Africa has the youngest population in the world. By 2050, one in four people globally will be African, and most will be young. This is either Africa's greatest asset or its greatest challenge—depending on whether we can create economic opportunities for this youth bulge.

Agriculture is where most of those opportunities will be. Not subsistence farming, but modern, profitable, technology-enabled agricultural enterprises.

Similarly, empowering women in agriculture isn't just about fairness—it's about unlocking productivity gains that could feed millions and lift entire communities out of poverty.

The new African farmer is young, tech-savvy, and entrepreneurial, and could increasingly include women. Our job is to clear the path for them.

Reflection Questions:

1. What would make agriculture attractive to young people in your community? What are the specific barriers they face?

2. How do gender dynamics in your context affect women's participation in agriculture? What one change would have the biggest impact?

3. If you were designing a programme to attract youth to agriculture, what would be its key features?

4. What role can digital technology play in making agriculture more accessible and profitable for young people and women?

Digital Innovation Spotlight: STRED Digital Farming and Development Institute

Scan the QR code to explore sample courses from SDFI, including "The Journey of a Tomato" and "Starting Your Agribusiness." Learn how mobile-based agricultural education is reaching thousands of young farmers across Africa.

The future of African agriculture walks on young legs and increasingly includes a woman's face. Our task is to ensure that the future is prosperous, dignified, and sustainable. The transformation of African agriculture depends on it.

Regional Integration and Global Markets

In 2019, I was part of the Expert Working Group (EWG) that conducted a preliminary assessment of Informal Cross-Border Trade ahead of the regional meeting. It was part of the reports delivered in the meeting on Enhanced quality of Informal Cross-Border Trade in ECOWAS region in Abidjan, Côte d'Ivoire (16-17 April, 2019).

The meeting was on the global review of aid for trade 2019 - regional session, with the theme "An Inclusive African Continental Free Trade Area: Aid for Trade and Empowerment of Women and Youth," organised by UNECA and the WTO, in cooperation with AfDB and others.

~

While on that preparatory mission, I still remember watching a truck loaded with cassava turn back at the Aflao border post between Ghana and Togo. The driver had all the right documentation from Ghana, but Togo's regulations were different. The cassava—perfectly good, desperately needed just 50 kilometres away—would rot before the bureaucratic tangle could be sorted out. Meanwhile, families on both sides of that invisible line were going hungry, and a farmer somewhere was losing money he'd counted on to pay school fees.

That scene has played out thousands of times across Africa's borders. But something remarkable is happening now. The African Continental Free Trade Area—AfCFTA—has the potential to rewrite this story entirely. For the first time in our history, we're not just talking about integration; we're actually building it.

This isn't just about trade agreements and tariff schedules, though those matter. It's about a Ghanaian farmer being able to sell her produce in Abidjan without losing half of it to border delays, and vice versa. It's about a Nigerian rice processor accessing quality seeds from Senegal. It's about treating Africa as one market of 1.3 billion people, not 54 fragmented ones. Operating in silos - more of the same - will take us nowhere.

The Promise and the Reality of AfCFTA

When African leaders signed the AfCFTA agreement in Kigali in 2018, the vision was breathtaking: a single continental market for goods and services, free movement of business travellers and investments, and a pathway to accelerate the establishment of a Continental Customs Union. For agriculture, the implications are transformative.

Consider the numbers for a moment. Africa currently imports about $50 - 78 billion worth of food annually. More recent data shows that this figure could rise significantly to between $90 billion and $110 billion by 2025 if no significant changes occur in food production and trade policies.

We should be growing and trading among ourselves. Our intra-African agricultural trade stands at a mere 18%, compared to 70% in Europe and 50% in Asia. We're exporting raw cocoa to Europe to make chocolate, then importing that chocolate back at ten times the price. We're selling raw cotton to Asia, then buying back the textiles. It's economic madness, and we all know it.

But here's what gives me hope: I've seen what happens when we get it right. Through my work for the ECOWAS Micro Reforms for African Agribusiness (MIRA) Project, I was hired by the Ministry of Food and Agriculture (MoFA), sponsored by AGRA on seed and fertiliser harmonisation. It was a relatively short assignment; however, given my role of developing a communication strategy for awareness creation and dissemination of the project's activities, I perceived at firsthand how removing regulatory barriers can unleash agricultural potential.

Align the seed rules—standards, certification, variety release—and suddenly farmers near borders can access more improved varieties. Yields can rise. Incomes can follow. It isn't magic; it's what happens when we treat colonially established borders as lines on a map, not walls around opportunity.

A Story From the Field: the Tomato Trader's Dilemma

Comfort is a tomato trader I met in Techiman, Ghana's largest food market. Every week, she used to watch perfectly good tomatoes from Burkina Faso arrive at the market, while her own stock from local farmers struggled to compete. Not because the Ghanaian tomatoes were inferior—they weren't. But because the regional supply chains were more efficient, better organised, seemed to be more responsive to consumers' needs, and benefited from cross-border cooperation that local value chains lacked.

"Dr Ayenor," she told me, "we're competing with one hand tied behind our backs. They have a specific seed variety, and a few of the traders have cold storage networks that cross borders. They have traders who know the regulations in countries and have found ways to get around regulatory barriers. We're still figuring out how to get our tomatoes from the farm to Accra without half of them spoiling."

That conversation changed how I think about regional integration. It's not about protecting our markets from regional competition—it's about learning from that competition and building the systems that make our own farmers competitive, both regionally and globally.

The Opportunities AfCFTA Creates for Agriculture

AfCFTA's significance for African farmers is tangible rather than theoretical. It's a practical framework that, if we implement it properly, will:

Expand markets dramatically: A Ghanaian maize farmer isn't just selling to Ghana's 31 million people anymore—she's potentially selling to 1.3 billion Africans. A Nigerian cassava processor (i.e., gari) can think beyond Lagos and consider Nairobi, Dakar, and Johannesburg.

Enable specialisation and efficiency: Not every country needs to grow everything. Kenya can focus on its excellent horticulture, Ghana on cocoa and cashew, Ethiopia on coffee, and Senegal on groundnuts. We can trade based on our comparative advantages, just like every other successful economic bloc does.

Attract investment in value chains: Investors love scale. A cashew processing plant that can serve multiple countries (Ghana and Côte d'Ivoire) is far more attractive than one limited to a single small market. AfCFTA makes regional value chains viable in ways they've never been before.

Strengthen our global negotiating position: When we negotiate with Europe, Asia, or America as a bloc of 1.3 billion people, we negotiate from strength, not weakness. We can demand fair prices for our commodities because we have alternatives. Thus, each other.

But Let's Talk About the Challenges

I've spent enough time in policy circles to know that signing agreements is the easy part. Implementation is where dreams go to die —unless we're brutally honest about the obstacles and determined to overcome them.

Infrastructure Gaps

You can eliminate every tariff in the world, but if it takes three days and five or more checkpoints to move goods from Accra to Lagos—a distance of just 500 kilometres—you haven't really created free trade. I've made that journey more times than I can count, and I've seen trucks lined up for kilometres at the border, drivers sleeping in their cabs, perishable goods slowly rotting in the tropical heat, and goods also deteriorating due to their direct exposure to intermittent rainfall during the journey.

We need cold chain infrastructure that crosses borders. We need roads that don't turn into rivers during the rainy season. We need ports that can handle increased trade volumes. We need digital systems that allow for tracking of goods and seamless customs clearance. These aren't luxuries—they're prerequisites for AfCFTA to work.

Regulatory Harmonisation

Here's a frustration I've lived with for years: a seed variety approved in Ghana might not be approved in Côte d'Ivoire, even though we share the same climate, soil types, and pest pressures. A fertiliser formulation certified in Nigeria might be rejected in Benin. Phytosanitary certificates issued in one country aren't automatically recognised in another.

The MIRA Project tried to address this for seeds and fertilisers within ECOWAS. Some progress has been made, but it was painfully slow. Every country's regulatory agency has its own procedures, its own forms, its own timelines. Harmonising all of this across fifty-four countries? That's the work of a generation, but we have to do it.

The Fear Factor

This might be uncomfortable: many African governments are terrified of AfCFTA, even if they won't admit it publicly. They're afraid their farmers can't compete. They're afraid of losing tariff revenues. They're afraid of powerful domestic interests that benefit from the current fragmented system.

I understand these fears—I've heard them in some closed-door ministerial and senior technical meetings. But here's what I always say: protection doesn't build competitiveness; competition does. Yes, we need transition periods. Yes, we need support to help farmers upgrade their practices. But hiding behind tariff walls hasn't made us prosperous in sixty years of independence. It's time to try something different.

Export Strategies That Actually Work

I've spent much of my career helping to develop export strategies for various commodities—cocoa, shea, cashew, mangoes, and more. I've seen strategies that worked brilliantly and others that failed spectacularly. Let me share what I've learned about what separates success from failure.

Quality Is Non-Negotiable

Ghana's cocoa commands a premium on world markets—not because we produce the most (Côte d'Ivoire produces more), but because we've maintained consistent quality for decades. The Ghana Cocoa Board's quality control system recently being led by my friend and classmate, Mr Frank Asante, has a great history here. Whatever may be its other flaws, Ghana Cocoa Board has protected our reputation.

I've evaluated export programmes in multiple countries, and the pattern is always the same: countries that maintain rigorous quality standards build lasting export markets. Those who chase volume at the expense of quality might see short-term gains, but they destroy their reputation and their premium.

This matters even more for intra-African trade. If Nigerian rice processors develop a reputation for consistent quality, Ghanaian consumers will seek out their products. If Kenyan horticulture maintains its high standards, restaurants across the continent will pay premium prices. Quality builds brands, and brands build sustainable markets.

Value Addition Is Where the Money Is

Africa produces 70% of the world's cocoa but captures only 2 - 6% of the $100 billion chocolate industry's value. We produce the coffee but don't roast it. We grow the cotton but don't weave it. We mine the gold but don't craft the jewellery.

AfCFTA gives us a chance to change this equation. With a continental market and bold leadership, it becomes viable to build processing facilities that serve multiple countries. A cocoa processing plant in Ghana can supply chocolate manufacturers in South Africa, Kenya, and Egypt. A textile mill in Ethiopia can serve fashion industries across the continent.

I've seen this work on a smaller scale. The shea value chain work I did for MEDA-FEAT, Canada showed how processing shea nuts into butter in Ghana, rather than exporting raw nuts, multiplied female pickers' incomes. Imagine scaling that across commodities and across the continent.

Market Intelligence and Adaptation

One of the most valuable things I learned from my work on international cocoa marketing—particularly organic and fair trade—is that markets are constantly evolving. What consumers wanted ten years ago isn't what they want today. What works in Europe might not work in Asia. What appeals to millennials differs from what their parents bought.

Successful exporters invest in understanding their markets. They attend trade fairs. They study consumer trends. They adapt their products to meet changing demands. They don't just grow what they've always grown and hope someone will buy it—they grow what markets are asking for. That's why I stated in Ghana's agricultural policies captured in the Coordinated Programme of Economic and Social Development (2017-2024) that agricultural commodity trading needed a paradigm shift. From supply-driven to demand-driven. That market has to drive the chain.

This applies to intra-African trade too. Urban African consumers are increasingly sophisticated. They want convenience, quality, and variety. They'll pay for products that meet their needs. But we have to understand those needs and respond to them accordingly.

Case Study: Ghana Cocoa's Quality Premium—Lessons for Other Commodities

Now, I share the story of how Ghana built and maintained its premium cocoa quality, because there are lessons here for every African commodity seeking to compete globally.

The Foundation: Consistent Standards

Ghana's cocoa quality didn't happen by accident. It's the result of a deliberate, decades-long system that touches every stage of the value chain:

- At the farm level, extension services (when they function properly) train farmers on proper fermentation and drying techniques. I've spent countless hours in cocoa communities, and I can tell you that farmers who follow these practices produce noticeably better beans.

- Licensed Buying Companies must meet quality standards before they can purchase cocoa. They can't just buy anything and dump it into the system.

- The Quality Control Division inspects and grades every single bag of cocoa before it's exported. Every. Single. Bag. It's labour-intensive and expensive, but it's why "Ghana cocoa" means something to chocolate manufacturers worldwide.

- The Cocoa Marketing Company negotiates forward contracts based on its reputation for quality, often securing prices above the world market average.

The Challenge: Maintaining Standards Under Pressure

Here's where it gets interesting—and difficult. When cocoa prices are high and farmers are getting good money, maintaining quality is relatively easy. Everyone's motivated. But when prices drop or farmers face urgent cash needs, the temptation to cut corners intensifies.

I've seen farmers try to sell under-fermented beans because they need money immediately and can't wait for the full fermentation period. I've seen beans dried on tarred roads instead of proper drying mats because they think it's faster. I've seen moisture content pushed to the maximum allowable limit (and sometimes beyond) because water weighs more than cocoa.

The system holds because the incentives are aligned. Farmers who consistently produce quality beans develop relationships with buying companies and get paid promptly. Those who try to game the system find their beans rejected. It's not perfect—no system is—but it works well enough to maintain Ghana's reputation.

The Opportunity: Extending the Model

This model can work for other commodities. During Shakti Pal's first fieldwork in Ghana on cashew, I was the TechnoServe staff member who worked with him closely in 2000-2001. I've seen elements of quality assurance principles succeed with cashew, shea, and even vegetables. The principles are universal:

Clear standards: Everyone knows what quality means and how to achieve it. I call it in my structured trading system hands-on training for value chain actors – "standards is the universal language between traders (buyers and sellers)".

Consistent enforcement: Standards aren't negotiable or subject to who you know.

Fair incentives: Quality is rewarded; poor quality is penalised.

Market feedback: Farmers understand how good agricultural practices affect quality and, in turn, how that affects prices and market access.

Institutional support: Extension services, research, and infrastructure enhance production quality.

Could we apply this to Kenyan avocados? Ethiopian coffee? Senegalese groundnuts? Absolutely. It requires investment, institutional commitment, and patience—but the payoff is sustainable premium markets that benefit farmers for generations.

Competing Globally While Feeding Ourselves

Here's a tension I've observed nations wrestled with throughout my career: How do we balance export agriculture with food security? How do we compete in global markets while ensuring our own people are fed?

Some people frame this as an either-or choice. They say, "Focus on exports to earn foreign exchange," or "Forget exports and grow food for Africans." But this is a false dichotomy. We need both, and AfCFTA helps us achieve both.

The Export-Food Security Nexus

Think about it this way: a farmer who earns good money from export crops can afford to buy food, invest in better farming practices, and send children to school. Export earnings at the national level finance infrastructure, healthcare, and education. The question isn't whether to export, but how to export in ways that support rather than undermine food security.

In the cocoa communities where I've worked, farmers who earn decent incomes from cocoa also maintain food crop farms. They use cocoa income to invest in their food production—buying better seeds, hiring labour for timely planting, and purchasing small-scale irrigation equipment.

The problems arise when export agriculture displaces food production entirely, when land tenure systems favour large-scale export farms over smallholder food producers, or when all the best land and resources go to export crops while food crops are relegated to marginal areas or lack institutional support.

AfCFTA's Role in Balancing the Equation

This is where AfCFTA becomes crucial: it makes food crops tradeable and valuable in ways they've never been before. A farmer in Ghana can now think of cassava not just as a subsistence crop but as a potential export to Nigeria's massive market. A maize farmer in Zambia can consider supplying deficit countries in Southern Africa. A rice producer in Senegal can target urban markets across West Africa.

When food crops become commercially viable, farmers invest in them. When farmers invest in them, productivity increases. When productivity increases, we produce surpluses that can be traded. It's a virtuous cycle that AfCFTA can trigger—if we get the implementation right.

This doesn't mean abandoning export crops like cocoa, coffee, or cotton. It means creating a more balanced agricultural economy where both export and food crops are profitable, where farmers can diversify their income sources, and where regional trade in food strengthens rather than threatens food security.

Building Regional Value Chains That Work

One of the most exciting aspects of AfCFTA is the potential for regional value chains—production systems that span multiple countries, each contributing what it does best. I've seen glimpses of this potential in my work across West Africa, and I'm convinced it's the future of African agriculture.

What Regional Value Chains Look Like

Imagine this: Senegal produces high-quality groundnuts. Ghana has processing capacity and technical expertise. Nigeria has the largest market in Africa. Instead of each country trying to do everything, they specialise and collaborate:

- Senegalese farmers focus on producing premium groundnuts, supported by research and extension services.

- Ghanaian processors import the raw groundnuts, process them into oil and paste, and add value through quality control and branding.

- Nigerian distributors and retailers market the finished products to their massive consumer base.

- Profits are shared across the chain, with each country benefiting from what it does best.

Elements of this already exist informally. AfCFTA provides the framework for formalising, scaling, and optimising these relationships.

The STRED Experience Across Borders

Through STRED's operations in Ghana and Nigeria, I've experienced firsthand the potential and challenges of cross-border agricultural collaboration. We've worked with farmer organisations in both countries, helped develop market linkages, and facilitated knowledge exchange.

What I've learned is that farmers on both sides of borders face remarkably similar challenges—and have remarkably similar solutions. A cassava processing technique that works in Ogun State, Nigeria, works equally well in the Volta Region, Ghana. A farmer organisation model that succeeds in Kaduna can be adapted for Tamale. We're not as different as our borders suggest.

But I've also learned how much those borders matter. Different currencies and languages complicate transactions. Different regulations create uncertainty. Different political systems mean different levels of government support. Moving equipment, inputs, or even knowledge across borders involves bureaucratic hurdles that would be comical if they weren't so costly.

AfCFTA won't eliminate these challenges overnight, but it provides the political framework and legal mechanisms to gradually reduce them. That's progress worth fighting for.

Keys to Successful Regional Value Chains

Based on my experience evaluating and supporting various cross-border agricultural initiatives, here's what makes regional value chains succeed:

Trust and relationships: Value chains are built on relationships between people, not just contracts between companies. Investing time in building trust across borders pays enormous dividends.

Clear value propositions: Every participant needs to understand what they gain from the collaboration. If benefits are one-sided, the chain will collapse.

Risk-sharing mechanisms: Cross-border trade involves currency risks, political risks, and market risks. Successful chains have mechanisms to share these risks fairly.

Quality standards: Everyone in the chain must commit to consistent quality. One weak link destroys the reputation of the entire chain.

Communication systems: Modern value chains require real-time communication about orders, quality, logistics, and payments. Digital tools make this possible in ways that weren't feasible even ten years ago.

Institutional support: Governments need to facilitate, not obstruct. This means harmonised regulations, efficient border procedures, and dispute resolution mechanisms.

The Digital Dimension of Regional Trade

I can't talk about regional integration without addressing the digital revolution that's making it possible in new ways. When I started my career, cross-border trade meant phone calls, faxes, and hoping your letter arrived. Today, a farmer in rural Ghana can connect with a buyer in Nairobi instantly, track shipments in real-time, and receive payments digitally.

The digital tools I discussed in Chapter 7 aren't just for connecting farmers to local markets—they're enabling regional and global trade in ways that were impossible before:

- Digital payment systems eliminate the currency exchange hassles that have plagued cross-border trade. A Nigerian buyer can pay a Ghanaian supplier instantly, with transparent exchange rates and low transaction costs.

- Blockchain-based traceability systems allow buyers anywhere in Africa (or the world) to verify the origin and quality of products. This builds trust and enables premium markets.

- Digital logistics platforms optimise cross-border transportation, reducing costs and delays. A truck that used to return empty from Nigeria to Ghana can now find cargo to return with through digital matching.

- E-commerce platforms connect African producers directly to African consumers, bypassing traditional intermediaries and their markups.

Through the STRED Digital Farming and Development Institute, we're training farmers not just in production techniques, but in digital tools for market access. Because in the AfCFTA era, digital literacy is as important as agricultural literacy.

What Needs to Happen Now

AfCFTA is signed. The vision is clear. But vision without implementation is just daydreaming. Based on my decades of experience in agricultural policy and trade facilitation, here's what needs to happen to make AfCFTA work for agriculture:

For Governments

- Accelerate or deepen implementation of the regulatory harmonisation, especially for seeds, fertilisers, pesticides, and phytosanitary standards. The ECOWAS MIRA Project showed this is possible—now scale it to the continental level.

- Invest in border infrastructure and procedures. Single-window systems for customs clearance. Joint border posts. Digital documentation.

- Resist the temptation to protect inefficient industries indefinitely. Use transition periods to help farmers become competitive, not to avoid competition forever.

- Establish dispute resolution mechanisms that work. When cross-border trade disputes arise (and they will), businesses need fast, fair, affordable ways to resolve them.

- Strengthen security on our ECOWAS and other regional trunk trading routes, rest stops and the borders against terrorism and armed robbery.

For Regional Economic Communities

- Lead by example. ECOWAS, EAC, SADC, and other regional blocs should serve as laboratories for testing and refining AfCFTA implementation.

- Share lessons learned. What works in East Africa might work in West Africa. What fails in Southern Africa can teach Central Africa what to avoid.

- Coordinate infrastructure development. Regional corridors and security, power grids, and digital networks need regional planning, not just national planning.

For the Private Sector

- Think regionally from day one. Don't build a business for Ghana or Nigeria—build it for West Africa. Don't plan for Kenya—plan for East Africa.

- Invest in quality and branding. Regional markets reward consistency and reputation. Build brands that Africans trust and seek out (i.e., Kasapreko).

- Collaborate across borders. The most successful African businesses in the AfCFTA era will be those that build cross-border partnerships, not those that try to do everything themselves.

For Farmers and Farmer Organisations

- Organise for scale. Individual farmers can't access regional markets effectively. Strong farmer organisations can negotiate contracts, ensure quality, and manage logistics.

- Invest in quality and consistency. Regional buyers will pay premiums to reliable suppliers who consistently meet specifications.

- Learn about markets beyond your borders. What do consumers in other African countries want? What standards do they expect? What prices will they pay?

- Embrace digital tools. They're not optional anymore—they're essential for accessing regional and global markets.

For Development Partners

- Support infrastructure development, especially cold chains, rural roads, and digital connectivity.

- Fund capacity building for farmers, businesses, and government officials on AfCFTA implementation.

- Support research on regional value chains, market opportunities, and best practices.

- Be patient. Regional integration is measured in decades, not project cycles. Commit for the long term.

A Personal Reflection on What's Possible

I've worked in agricultural development across West Africa for over thirty years. I've seen initiatives that promised transformation but delivered disappointment. I've watched as grand plans collapsed under the weight of poor implementation, political interference, or simple lack of follow-through.

So, I approach AfCFTA with both hope and realism. Hope, because the potential is genuinely transformative. Realism, because I know how hard implementation will be.

But here's what keeps me optimistic: I've also seen what's possible when we get things right. I've seen the potential of ECOWAS seed harmonisation to reduce barriers and increase farmers' access to better varieties. I've seen Ghana's cocoa quality system maintain premium markets for decades. I've seen farmer organisations like Kuapa Kokoo build successful businesses that benefit thousands of smallholders. I've seen digital tools connect farmers to markets in ways that seemed like science fiction when I started my career.

AfCFTA isn't guaranteed to succeed. But it's our best chance in generations to fundamentally reshape African agriculture and African economies. It's our opportunity to trade with each other, to build value chains that keep wealth on the continent, to create markets large enough to justify the investments we need.

The question isn't whether AfCFTA is perfect—it isn't. The question is whether we'll commit to making it work, whether we'll push through the inevitable obstacles, and whether we'll maintain focus when implementation gets difficult.

I believe we will. I've seen too much potential, worked with too many capable people, and witnessed too many small successes to be pessimistic about our collective future.

That truck of cassava I saw turned back at the Aflao border? In the AfCFTA future, it crosses smoothly, the cassava reaches its market fresh, the farmer gets paid fairly, and families on both sides of the border eat well. That's an achievable goal if we commit to the work.

The fields can become fortunes—not just for individual farmers, but for entire regions, for the continent. But only if we build the systems, the infrastructure, the relationships, and the trust that make regional integration real.

The opportunity is ours. The time is now. Let's not waste it.

Reflection Questions

1. What products from your country could find markets in neighbouring countries if trade barriers were reduced? What would it take to make those products competitive?

2. How can your country balance export agriculture with food security? Are these goals necessarily in conflict, or can they be complementary?

3. What lessons from Ghana's cocoa quality system could apply to other commodities in your context?

4. What are the biggest obstacles to regional agricultural trade in your region? Which of these obstacles could be addressed relatively quickly, and which require long-term commitment?

5. How can digital tools facilitate regional trade in your context? What infrastructure or capacity building would be needed?

6. What role should farmer organisations play in accessing regional markets? How can they be strengthened to fulfil this role?

7. How can STRED Consult, STRED Digital Farming and Development Institute (SDFI), and other partners help you address agricultural or rural research and development challenges you may face as government, private sector, development partners, or donors?

Chapter 12

From Vision to Action - Your Next Steps

We've covered a lot of ground together in this book. We've walked through cocoa farms in Ghana's Eastern Region, visited cassava processing centres in Nigeria, explored digital farming initiatives, examined policy frameworks, and imagined what regional integration could mean for African agriculture.

We've talked about soil health and market access, about youth and women, about technology and tradition. But here's the question that matters most: What happens now? What do you—yes, you, the person reading these words—do on Monday morning or from now onward?

Because that's the test of any book about development, isn't it? Does it gather dust on a shelf, or does it spark action? Does it add to the endless pile of reports and analyses about what Africa could be, or does it contribute to what Africa is becoming? Can this book contribute to the practical strategy that Africa needs to feed itself and the globe?

For over 30 years in agricultural development, I've written policy documents that were never implemented, designed programmes that never launched, and attended conferences where we talked brilliantly about transformation and then went home to business as usual. I've also been part of initiatives that genuinely changed lives—that helped farmers double their incomes, that connected young people to agricultural opportunities, that built systems that outlasted the projects that created them.

The difference between the two? ACTION. Specific, concrete, persistent action by people who refused to accept that things had to stay the way they'd always been.

This final chapter is about moving from vision to action. It's about what you can do, whoever you are, wherever you are, to be part of Africa's agricultural transformation. Because transformation isn't something that happens to us—it's something we create, one decision, one investment, one innovation at a time.

If You're a Farmer

You are the foundation of everything we've discussed in this book. Without you, there is no agricultural transformation. The question is: how do you move from where you are now to where you want to be?

Start With One Thing

I know the challenges you face can feel overwhelming. Poor soil. Unreliable rain. Pests and diseases. Low prices. Lack of credit. Inadequate storage. The list goes on. But here's what I've learned from working with thousands of farmers: you don't have to solve everything at once. Start with one thing.

Maybe it's improving your soil. Start composting. Plant cover crops. Try the techniques we discussed in Chapter 4. You don't need to transform your entire farm overnight—start with one plot, see what works, learn from it, then expand.

Maybe it's improving your market access. Join a farmer organisation if there's a good one in your area. If there isn't, start one with a few neighbours you trust. Collective marketing gives you negotiating and countervailing power that individual farmers will never have.

Maybe it's trying one new crop or variety. Something that's suited to your climate but might fetch a better price. Something that diversifies your income so you're not dependent on a single crop's success or failure.

Maybe it's adopting one digital tool. If you have a mobile phone—and most of you do—there's probably an app that can help you access weather information, market prices, or agronomic advice. Start with one. Learn how to use it. See if it helps. If it does, explore others.

The point is this: don't let the size of the challenge paralyse you. Pick one thing you can control, one thing you can improve, and start there. Small changes compound over time.

Invest in Learning

The most successful farmers I know are perpetual learners (i.e., Mr Duncan Yeboah, Eric Nyamekye, Mr Maxwell Akanadem and Evelyn Agyenim Boateng). They attend training sessions when they're offered. They visit demonstration plots. They observe what their neighbours are doing and adapt the good ideas. They ask questions. They experiment.

Agriculture is changing faster now than at any point in human history. New varieties, new techniques, new technologies, new markets— they're all emerging constantly. The farmers who thrive are those who stay curious and keep learning.

If there's an extension service in your area, use it. Yes, I know extension services are often understaffed and underfunded. I know the advice isn't always relevant or timely. But take what's useful and leave the rest.

If there are farmer field schools or demonstration plots, participate. The best learning happens when you can see techniques in action, try them yourself, and discuss results with other farmers.

If there are digital learning resources available—videos, radio programmes, SMS services—use them. The STRED Digital Farming and Development Institute and similar initiatives are creating content specifically designed for African farmers. It's not perfect, but it's getting better all the time.

Think Like a Business Owner

I know many of you think of yourselves as farmers, not business owners. But here's the truth: if you're selling anything you produce, you're running a business. And the more you think like a business owner, the more successful you'll be.

What does that mean practically?

Keep records. *I* know, I know—you're busy, and record-keeping feels like bureaucratic nonsense. But how can you know if you're making money if you don't track your costs and revenues? It doesn't have to be complicated. A simple notebook where you record what you spend on inputs and what you earn from sales will tell you which crops are profitable, and which aren't.

Calculate your costs. Include everything—seeds, fertiliser, labour (including your own), land preparation, harvesting, transport. Many farmers think they're making money when they're actually losing it because they don't account for all their costs, especially their own labour.

Know your market. Who are you selling to? What do they want? What will they pay? Can you sell directly to consumers instead of through middlemen? Can you add value through processing or better packaging? These aren't abstract questions—they're the difference between poverty and prosperity.

Plan ahead. What will you plant next season? Where will you get your inputs? How will you finance the season? Where will you sell your harvest? Farmers who plan ahead consistently do better than those who make it up as they go.

Don't Do It Alone

Individual smallholder farmers are at a massive disadvantage in modern agricultural markets. You have no negotiating power with buyers. You can't afford the inputs or equipment that would make you more productive. You can't access the credit or insurance that would reduce your risk.

But organised farmers can do all of these things. This is why I keep emphasising farmer organisations. Yes, many of them are poorly run. Yes, some are captured by elites who benefit more than ordinary members. But when they work well—and I've seen and helped develop many that do— with a proper governance system, they transform what's possible for smallholder farmers.

I repeat, if there's a good farmer organisation in your area, join it. If there isn't, consider starting one with farmers you trust. Start small— maybe just five or ten of you who agree to market collectively or share equipment. Build trust. Demonstrate value. Grow gradually.

The most successful farmer organisations I've worked with share certain characteristics:

- They're democratic and transparent

- They provide clear, tangible benefits to members

- They're well-managed, with proper record-keeping and financial controls

- They maintain quality standards that benefit everyone

- They invest in their members' capacity through training and information sharing

You can build this in your community. It won't be easy, but it's possible.

Take Care of Your Land

I can't emphasise this enough: your land is your most valuable asset. Don't look at immediate transient benefits and invite illegal miners to destroy your land. If you mine its fertility without replenishing it, you're destroying your own future. Everything we discussed in Chapter 6 about soil health isn't optional—it's essential for your long-term survival as a farmer.

I've seen too many farmers chase short-term yields at the expense of long-term sustainability. I've seen land degraded by illegal surface miners to the point where it can barely grow anything. I've seen farmers forced to abandon their land and migrate to cities because they destroyed the resource that sustained them.

Don't let this be your story. Invest in your soil. Protect your water sources. Maintain the trees on your farm. Practice integrated pest management instead of relying solely on chemicals. Think about what you're leaving for your children.

Sustainable farming isn't just good for the environment—it's good for your income. Healthy soil produces better yields with fewer inputs. Diverse farms are more resilient to pests, diseases, and climate shocks. Farmers who take care of their land consistently outperform those who don't, year after year.

If You're a Young Person Considering Agriculture

I know what many of you are thinking: "Agriculture? That's what my parents do. That's what I'm trying to escape." I understand. I've heard it countless times. You've seen your parents work incredibly hard for very little return. You've seen farming associated with poverty, backwardness, and lack of opportunity.

But I'm asking you to look again. Because agriculture in Africa is changing, and the opportunities for young people who approach it with entrepreneurial mindsets and modern tools are genuinely exciting. Our parents, who were farmers, didn't have the opportunities we have today.

Agriculture Doesn't Mean Subsistence Farming

When I talk about agricultural opportunities for youth, I'm not suggesting you should do exactly what your parents did. I'm talking about approaching agriculture as a business, using technology, adding value, and capturing more of the value chain.

Maybe you're not farming at all—maybe you're providing services to farmers. Equipment rental. Input supply. Soil testing. Drone services for crop monitoring. Digital platforms for market information. Transportation and logistics. Storage and warehousing. Processing and packaging.

Maybe you're farming, but you're doing it differently. You're using drip irrigation and growing high-value vegetables for urban markets. You're raising poultry or fish using modern techniques. You're growing organic produce for export. You're cultivating mushrooms, snails, or other non-traditional products that have strong market demand.

Maybe you're building the digital tools that other farmers need. Apps for market information, weather forecasting, pest identification, and financial services. If you have tech skills, agriculture needs you desperately.

Maybe you're in the processing and value addition space. Taking raw agricultural products and turning them into something consumers want to buy. Cassava into flour or starch. Tomatoes into paste. Fruits into juice or dried snacks. Shea nuts into cosmetic butter.

The point is: there are dozens of ways to build a career in agriculture that don't look like traditional subsistence farming. And many of them are more profitable than the urban jobs you're chasing.

What You Bring That's Valuable

You have advantages that older farmers don't. You're comfortable with technology. You're not wedded to "how we've always done it." You're willing to experiment and take risks. You can see opportunities that others miss.

I've met young agripreneurs across West Africa who are building impressive businesses:

- A young woman in Ghana who built a thriving business processing cassava into high-quality gari for urban markets

- A young man in Nigeria who intends to provide drone services to large farms for crop monitoring and spraying

- A group of university graduates who started an organic vegetable farm supplying high-end restaurants and supermarkets

- Tech-savvy entrepreneurs such as Francis and Michael of AgroCenta, who built a platform connecting smallholder farmers directly to urban consumers

None of them is getting rich quick. All of them worked hard, made mistakes, and learned as they went. But all of them are building sustainable businesses that provide good incomes and create employment for others.

You can do this too. But you need to approach it seriously, as a business, not as a fallback option when nothing else works out.

What You Need To Succeed

Skills and knowledge: You can't just jump into agriculture because you're young and enthusiastic. You need to understand the technical aspects of whatever you're doing. Take courses. Find mentors. Learn from successful agripreneurs. The STRED Digital Farming and Development Institute and similar programmes offer training specifically designed for young people entering agriculture.

Start-up capital: Yes, you need money to start. But probably less than you think. Many successful agripreneurs started small—very small—and grew gradually. Look for youth agricultural financing programmes. Consider partnerships where you provide labour and management while someone else provides land or capital. Explore value chain financing where input suppliers or buyers provide credit, and be honest and trustworthy.

Market knowledge: Before you produce anything, understand who will buy it, at what price, and what quality standards they expect. Too many young farmers produce first and then struggle to find markets. Do it the other way around—identify the market opportunity, then figure out how to supply it. Market-driven, not supply-driven.

Persistence: I won't lie to you—agriculture is hard. You'll face challenges you didn't anticipate. Crops will fail. Markets will disappoint. Partners will let you down. The difference between those who succeed and those who quit is persistence. Learn from failures. Adapt. Keep going. Winners don't quit.

Networks: Connect with other young agripreneurs. Join youth farmer associations. Attend agricultural events and competitions. The relationships you build will be as valuable as anything else—for sharing information, solving problems, and creating opportunities.

A Challenge to You

Africa's agricultural transformation will not happen without you. Your parents' generation has done what they could with the tools and knowledge they had. But the next phase—the digital revolution, the value addition, the regional integration, the climate adaptation—that's your generation's work.

I'm not romanticising agriculture. I'm not pretending it's easy. But I am saying that if you're looking for opportunities to build something meaningful, to create wealth, to make a difference in your community, agriculture offers possibilities that few other sectors can match.

The question is: are you willing to look past the stereotypes and see the opportunities? Are you willing to work hard and smart? Are you willing to be part of the solution instead of just complaining about the problems and having no jobs?

If the answer is yes, agriculture needs you. And you might be surprised by what agriculture can offer you in return.

If You're a Woman in Agriculture

You are already the backbone of African agriculture. You produce or process most of the food. You do most of the processing and marketing. You manage household food security. But you may do all of this with one hand tied behind your back—with less access to land, credit, inputs, training, and markets than your male counterparts.

Everything I said to farmers generally applies to you, but you face additional barriers that need to be named and addressed. However, things are gradually changing – though slowly.

Know Your Rights

In many African countries, laws have changed to give women more rights to land ownership and inheritance. But laws on paper don't always translate to reality on the ground. Customary practices may often still favour men.

Know what the law says in your country. If you have legal rights to land, insist on them. Get documentation—title deeds, rental agreements, whatever proves your right to use the land you're farming. Without secure land rights, you can't make long-term investments or access credit.

If customary practices in your community deny women land rights, work with other women to challenge them. I've seen women's groups successfully advocate for change in their communities. It's not easy, but it's possible.

Access the Resources You Need

Women-focused agricultural programmes exist in most African countries—financing programmes, training initiatives, and input subsidies. They're not always well-publicised, and they're often underutilised. Find out what's available in your area and access it.

Join women's farmer groups. They provide solidarity, shared learning, and often better access to resources and markets than you'd have individually. The most successful women farmers I know are almost always part of strong women's organisations.

Don't accept being excluded from training or extension services. If extension agents only talk to men in your household, insist on being included. If training sessions are scheduled at times when women can't attend, demand that they be rescheduled. Your voice matters.

Build on Your Strengths

Women often excel at aspects of agriculture that are increasingly valuable: attention to quality, relationship-building, diversification, food-processing activities, and market knowledge. These aren't soft skills—they're competitive advantages.

I've seen women's groups maintain quality standards that male-dominated groups couldn't match, because the women understood that their reputation was their most valuable asset. I've seen women farmers diversify their production in ways that made them more resilient to shocks. In 2012, while working with Mark Lewis, Senior Agricultural Specialist for IFC, in the Democratic Republic of Congo, I saw women traders build market relationships that enabled them to secure better prices than men did.

Don't let anyone tell you that being a woman is a disadvantage in agriculture. Yes, you face barriers. But you also bring strengths that are increasingly recognised and valued.

Support Other Women

As you succeed, help other women succeed. Mentor younger women. Share information. Create opportunities for other women in your value chain. Advocate for policies that support women in agriculture. Dr Mei Zegers emphasised these values when we were working together in the field in the Ashanti region, under the European Union-sponsored Archipelago programme TVET initiative with Kwadaso female trainees.

The most inspiring women agricultural leaders I know are those who use their success to lift others. They don't just break through barriers themselves—they work to remove the barriers so other women don't have to fight the same battles.

If You're in Government

You have enormous power to either enable or obstruct agricultural transformation. The policies you make, the budgets you allocate, the regulations you enforce—they shape what's possible for millions of farmers.

In working with government officials across West Africa for three decades, I've seen brilliant, committed public servants who genuinely want to serve their countries. I've also seen incompetence, corruption, and indifference.

Get the Basics Right

Before you launch another ambitious programme or create another new agency, make sure the basics are working.

Extension services: Are farmers actually receiving useful, timely advice? Or is your extension system understaffed, underfunded, and irrelevant? Fix this before you do anything else.

Input supply: Can farmers access quality seeds, fertilisers, and other inputs nearer when they need them, at prices they can afford? Or is your input subsidy programme riddled with corruption and inefficiency?

Infrastructure: Can farmers get their produce to market without half of it spoiling? Or are rural roads impassable during the rainy season? Is there storage infrastructure? Cold chain facilities? Processing capacity?

Research: Is your agricultural research system generating technologies that farmers can actually use? Or is it disconnected from farmers' realities and priorities?

Market information: Do farmers know what prices they should expect? Or are they at the mercy of middlemen because they have no information?

Create an Enabling Environment

Your job is to create conditions where farmers and agricultural businesses can thrive. That means:

Stable policies: Stop changing agricultural policies with every new minister or administration. Farmers need predictability to make long-term investments.

Fair regulations: Regulations should protect consumers and the environment without strangling businesses with bureaucracy. Review your regulatory frameworks honestly—are they serving their intended purpose, or are they just creating opportunities for rent-seeking?

Investment in public goods: Infrastructure, research, extension, education—these are things markets won't provide adequately on their own. This is where government investment is essential and appropriate.

Level playing field: Don't favour large farms over small ones, or men over women, or politically connected businesses over others. Create conditions where everyone can compete fairly as citizens.

Implement What You've Committed to

I've sat through countless policy launches where ministers gave inspiring speeches about agricultural transformation. I've read beautiful policy documents with ambitious targets. And then I've watched as nothing happened—no budget allocation, no implementation plan, no follow-through.

If you're going to announce a policy, implement it. If you can't implement it, don't announce it. Your credibility—and your country's—depends on following through on commitments. One can only hope that the current 24-hour economic policies do not become another rhetoric or only slogans, but are effectively implemented.

This applies especially to AfCFTA. Your country has signed the agreement. Now implement it. Harmonise your regulations with neighbouring countries. Invest in border infrastructure. Remove non-tariff barriers. Stop protecting inefficient industries indefinitely.

Listen to Farmers

When was the last time you actually talked to farmers? Not at a staged event where they tell you what you want to hear, but real conversations where they tell you what's actually happening in their lives? The best agricultural policies I've seen were developed with genuine input from farmers. Where practices on the ground informed theories and vice versa. The worst were designed by technocrats in capital cities who hadn't set foot on a farm in years.

Create mechanisms for regular farmer feedback. Support farmer organisations as genuine partners, not just as implementers of your programmes. When farmers tell you something isn't working, listen to them.

Measure What Matters

Stop measuring success by how much money you spent or how many farmers you "reached." Measure outcomes: Are farmers' incomes increasing? Is productivity improving? Is food security getting better? Is the sector creating jobs for young people?

Be honest about what's working and what isn't. If a programme isn't delivering results, have the courage to change it or end it, rather than continue funding failure because admitting mistakes is politically uncomfortable.

If You're in the Private Sector

You have a crucial role in agricultural transformation. You provide the inputs, the equipment, the processing capacity, and the market access that farmers need. You create jobs. You drive innovation. You take risks that some governments can't or won't take.

But let's be honest: too much private sector involvement in African agriculture has been extractive rather than developmental. Taking value out of rural communities rather than creating it. Maximising short-term profits rather than building sustainable relationships.

It doesn't have to be this way. The most successful agricultural businesses I've seen are those that genuinely partner with farmers, that invest in their suppliers' success, that take a long-term view. Akanadem Farms and its 2,000+ smallholder farmers in Sandema, Chichuliga and its environs in Upper East region of Ghana are some of the few good examples I have witnessed.

Build Real Partnerships With Farmers

Contract farming can be exploitative, or it can be transformative—the difference is in how you structure the relationship.

If you're just using contracts to lock farmers into selling to you at low prices while bearing all the risk themselves, you're extracting value, not creating it. Farmers will resent you, they'll try to side-sell when they can, and your supply chain will be unreliable.

But if you structure contracts that share risks and rewards fairly, that provide farmers with inputs and technical support, that pay promptly and transparently, you'll build trust, loyalty and reliability that benefits everyone. Of course, there are a few unscrupulous farmers as well. The best contract farming arrangements I've evaluated include:

- Fair pricing mechanisms that reflect market conditions

- Input provision on credit, recovered at harvest

- Technical support to help farmers meet quality standards

- Prompt payment

- Long-term relationships, not one-season transactions

It's good business—farmers who trust you and benefit from working with you are reliable suppliers. Those who feel exploited are not.

Invest in Value Addition

Stop just buying raw materials and exporting them. Build processing capacity in Africa, such as Yedent Processing in Sunyani. Create jobs. Capture more of the value chain.

Yes, there are challenges—infrastructure gaps, skills shortages, irregular supply of raw materials and regulatory hurdles. But these are solvable problems, and the opportunities are enormous. AfCFTA creates a continental market that makes processing investments viable in ways they've never been before.

I've seen successful processing businesses across West Africa—cassava processing, shea butter production, fruit juice manufacturing, and cocoa processing. They're profitable, they create employment, and they benefit farmers by providing reliable markets and often better prices than export markets for raw materials.

Think Long-Term

I know you may face pressure for quarterly results. I know investors want quick returns. But agriculture doesn't work on quarterly cycles. Crops take seasons to grow. Relationships take years to build. Sustainable businesses take time to develop.

The most successful agricultural businesses I've seen are those that take a long-term view. They invest in their supply chains. They build relationships with farming communities. They develop brands and reputations that take years to establish but provide competitive advantages that last.

If you're just looking for quick profits, agriculture probably isn't for you. But if you're willing to invest for the long term, the opportunities are substantial. Sadly, many banks in Ghana, including some development ones—only look at agriculture to meet donors' reporting needs, avoid public criticism or please politicians.

Use Your Influence Responsibly

You have influence with governments. You have resources. You have platforms. Use them to advocate for policies that benefit the sector as a whole, not just your interests.

Support regulatory harmonisation. Advocate for infrastructure investment. Push for transparent, competitive markets. Oppose corruption and rent-seeking, even when you might benefit from it in the short term.

The most respected business leaders I know are those who use their influence to strengthen the entire agricultural ecosystem, not just their own position within it. And there are very few.

If You're a Development Partner

Suffice to say that there are a very few major international development partners, donors and NGOs that I've not worked or collaborated with in my career.

I mentioned earlier that I've seen projects that genuinely transformed communities and others that wasted millions of dollars, leaving nothing behind but abstract reports.

Most mean well. I don't doubt that. But good intentions aren't enough. After decades of agricultural and international development projects, Africa is still importing food it could grow, farmers are still poor, and rural youth are still fleeing to cities. Something isn't working.

Listen More, Prescribe Less

You don't have all the answers. The fact that something worked in Asia or Latin America doesn't mean it will necessarily work in Africa. The fact that your theory of change looks good on paper doesn't mean it reflects reality on the ground. Spend more time in diagnostic studies. Listen to the institutions, but also pay attention to the real practitioners on the ground

Spend more time listening to farmers, to local organisations, to government officials who actually implement programmes. They know things you don't. They understand contexts you don't. They can tell you why your brilliant idea won't work—if you're willing to listen.

The best development projects I've been part of were those in which external partners brought resources and technical expertise, while local actors shaped the approach based on their knowledge of what would actually work in their context.

Support Systems, Not Just Projects

Stop funding 3-year projects that create temporary structures that collapse the moment your funding ends. Support the development of permanent systems—government extension services, farmer organisations, market infrastructure, and research capacity.

Yes, this is harder. Yes, it takes longer and perhaps slower. Yes, it's less satisfying than launching a new project with your logo on it. But it's what actually creates lasting change. When you fund a project, ask yourself: what will remain when our funding ends? If the answer is "nothing," redesign the project.

Align With Government Priorities

I know government priorities aren't always what you think they should be. I know government systems are often inefficient. But if you're working around government rather than with it, you're not building sustainable capacity. USAID's TIPCEE project was one of the best projects. Yet, at the institutional level, very limited mainstreaming, particularly vertical integration (Upscaling), took place. One unfortunate example is that despite its great achievements in the horticultural sector, it didn't find a successful way to work much more closely with MoFA's horticultural department.

Lessons learnt: support government systems, even when it's frustrating. Help strengthen them rather than creating parallel structures. Align your projects with national strategies. Work through government channels when possible.

Yes, this means your money might not be spent as efficiently as if you controlled everything directly. But it means you're building capacity that will outlast your project.

Measure Real Impact

Stop measuring success by outputs—how many farmers trained, how many demonstrations conducted, how much money spent. Measure outcomes—are farmers' incomes actually increasing? Is food security improving? Are the systems you supported still functioning after your project ends?

Be honest about what's working and what isn't. Share your failures as openly as your successes, so others can learn from them. The development sector's reluctance to admit failure means we keep repeating the same mistakes. And alleviating our own poverty as development practitioners, rather than that of the more vulnerable rural folks and farmers. Let's be honest about it, despite our unquestionable collective impacts.

Commit for the Long Term

Agricultural transformation takes decades, not project cycles. If you're not willing to commit to the long term, be honest about that and design your interventions accordingly.

The most successful development partnerships I've seen are those in which partners are committed to a country or region for decades, building relationships, learning what works, adapting their approaches, and staying engaged through political changes and economic cycles.

If You're a Researcher or Academic

You have knowledge and analytical capacity that the sector desperately needs. But too much agricultural research in Africa is disconnected from farmers' realities and priorities. The career paths favour a number of academic or journal publications over real impact on Ghanaian and African lives in our communities.

Research What Matters to Farmers

Before you design your research, talk to farmers. What problems are they facing? What questions do they need answered? What would actually make a difference in their lives? Conduct diagnostic studies with farmers and stakeholders, including scientists – not just desktop literature reviews.

I've seen a lot of research that's intellectually interesting but practically irrelevant. Research that answers questions no one is asking. Research that's published in journals that literate farmers will never read or benefit from, and policymakers will never see. Hence, the unfortunate huge disconnect between research and policy implementation.

The best agricultural research I've encountered is that which emerges from genuine engagement with farmers' problems and produces results that farmers and other commodity value chain actors can and actually use.

Make Your Research Accessible

If your research findings are locked behind paywalls in academic journals, written in jargon that only other academics understand, they're not contributing to agricultural transformation.

Publish in open-access formats. Write policy briefs and action research papers for government officials. Create extension materials for farmers. Give presentations to farmer organisations. Use social media and digital platforms to share your findings widely. Your research is funded by public money or donor funds meant to support development. You have an obligation to make it accessible to those it's meant to benefit.

Engage in Implementation

Don't just publish and move on to the next research project. Stay engaged with how your research is being used (or not used). Work with extension services to translate your findings into practical recommendations. Partner with farmer organisations to test and adapt your recommendations in real-world conditions.

The most impactful researchers I know are those who see themselves as partners or learners in agricultural development, not just as observers and analysts.

If You're a Consumer

Yes, you. You have more power than you probably realise. Every time you buy food, you're making choices that affect farmers, value chains, and agricultural systems.

Buy Local When You Can

I'm not suggesting you should never buy imported food. But when local alternatives are available and of good quality, choose them. Your purchases support local farmers, create local jobs, and build local value chains.

This is especially important for processed foods. That fruit juice made from local fruits, that cassava bread from flour made with local cassava, that vegetable oil from local groundnuts—buying them supports the value addition and job creation we've been talking about throughout this book.

Demand Quality, but Pay for It

African farmers can produce high-quality products. But quality costs more to produce. If you always choose the cheapest option regardless of quality, you're incentivising a race to the bottom.

Be willing to pay a bit more for products that are higher quality, more sustainably produced, or certified as fair trade or organic. Your willingness to pay for quality creates markets that reward farmers for producing quality.

Reduce Food Waste

Africa loses about 30% of its food production to post-harvest losses and waste. Some of this happens in the value chain, but a lot happens in our homes—food we buy and then throw away.

Plan your purchases. Store food properly. Use what you buy. When you waste food, you're wasting the resources that went into producing it and the income that farmers could have earned from it.

Support Farmers' Markets and Direct Sales

When possible, buy directly from farmers or through farmers' markets; more of your money goes to the farmer rather than to middlemen. You get fresher food, farmers get better prices, and you build relationships with the people who grow your food. In rural communities and smaller towns, with time, farmers will bring some of their produce to your doorstep. You can even use mobile phones to place orders as well.

Some African cities now have farmers' markets or direct-sales initiatives. Support them. They're building the kind of direct farmer-consumer connections that benefit everyone. Strengthening farmer cooperatives or organisations under the 24-hour economy could promote this idea better.

A Final Word: The Transformation We Need

I started this book with a vision: African farmers prospering, rural communities thriving, young people seeing agriculture as an opportunity rather than drudgery, Africa feeding itself and exporting surpluses to the world. I am certainly not the first to envision these laudable ideas. The question is how many of us have documented our experiences in such simple language for our people? In Africa, we are all guilty of not writing enough. This is an invitation to many colleagues to equally share their experiences as well, for us to learn even more.

This vision is achievable. Not easily. Not quickly. But achievable.

I've seen enough successes—small and large—to know that transformation is possible. I've worked with farmers who've tripled their incomes. I've seen farmer organisations that have built successful businesses (i.e., Bodom Cooperative farmers, Akrofrom Catholic

Women Gari Processing Association, still operating on the Wenchi –
Techiman roadside). I've watched young agripreneurs create thriving
enterprises. I've participated in policy changes that genuinely
improved farmers' lives. I've witnessed digital tools connect farmers to
markets in ways that seemed impossible a decade ago.

But I've also seen enough failures to know that transformation isn't a
walk in a park. It requires action—specific, sustained, coordinated
action by all of us.

It requires farmers willing to try new approaches and invest in their
land. It requires young people willing to see agriculture's potential. It
requires women demanding their rightful place in agricultural systems.
It requires governments creating enabling environments rather than
obstacles. It requires private sector partners who build rather than
extract. It requires development partners who support systems rather
than just projects. It requires researchers who engage with real-world
problems. It requires consumers who make conscious choices. It also
requires consumers making conscious choices to buy and consume
locally.

It requires all of us.

I've given you my best thinking and laced with my own experiences in
this book—lessons from three decades of work, especially in Ghana
and across Africa, insights from successes and failures, practical
recommendations for action. But this book is just words on paper
unless you do something with it.

So, here's my challenge to you: Don't just read this book and put it on a
shelf. Don't just nod along and then go back to business as usual. Pick
one thing—one specific action you can take based on what you've read
—and do it.

If you're a farmer, maybe it's joining a farmer organisation or trying one
new soil improvement technique.

If you're a young person, maybe it's seriously investigating one
agricultural business opportunity.

If you're in government, maybe it's fixing one broken system or implementing one policy you've been ignoring.

If you're in the private sector, maybe it's restructuring one contract to be fairer to farmers or investing in one processing facility.

If you're a development partner, maybe it's redesigning one project to support systems rather than create parallel structures.

If you're a researcher, maybe it's engaging with one farmer organisation to make your research more relevant.

If you're a consumer, maybe it's consciously choosing local products or reducing your food waste.

One action. Then another. Then another.

That's how transformation happens—not through grand declarations or ambitious plans, but through thousands of people taking concrete actions, learning from them, and taking more actions.

The fields can become fortunes. But only if we make it so. We need to go beyond well-crafted and soundbites from campaign slogans. We need concrete learning and action.

The opportunity is ours. The time is now.

What will you do?

Your Personal Action Plan

Take a few minutes right now—not later, now—to write down your answers to these questions:

1. What is one specific thing I can do in the next week based on what I've learned from this book?

2. What is one specific thing I can do in the next month?

3. What is one specific thing I can do in the next year?

4. Who can I partner with or learn from to increase my chances of success?

5. How will I measure whether my actions are making a difference?

6. Who will I share this book with, and what conversation will I start with them about agricultural transformation?

Write these down. Put them somewhere you'll see them. Share them with someone who will hold you accountable. And then—most importantly—act on them. Africa needs action, not speeches!

A Personal Note

When I started my career in agricultural development over thirty years ago, I was young, idealistic, and convinced that transformation would happen quickly if we just had the right policies and programmes.

I've learned since then that transformation is slower, harder, and more complex than I imagined. I've experienced frustrations, disappointments, and failures. I've seen initiatives I believed in collapse. I've watched as political interference, institutional politics, bad corporate governance, and conflicts of interest have undermined good programmes. I've witnessed corruption waste resources that could have changed lives. Lost opportunities due to a few greedy people all over our dear continent.

But I've also seen enough success to remain hopeful. I've worked with farmers whose lives genuinely improved. I've been part of policy changes that made a real difference. I've watched young people build impressive agricultural enterprises. I've seen digital tools transform what's possible. I've witnessed farmer organisations become powerful economic actors (KuapaKoko, Cocoa Organic Farmers' Association, Nsawkaw, Kabile, and Sampa Cashew Farmers' Cooperatives).

And I've learned that transformation doesn't happen through any single intervention or any one actor. It happens through the accumulated efforts of thousands of people, each doing what they can with the resources and opportunities they have.

You're one of those people. What you do matters. Your actions—however small they might seem—contribute to the larger transformation we all want to see.

I've given you everything I know and have experienced in this book. Now it's up to you.

Let's turn these fields into fortunes—*together*.

About the Author

Dr Godwin Kojo Ayenor is one of West Africa's most experienced and respected agricultural and rural development practitioners, with over three decades of hands-on work transforming the lives of smallholder farmers across the region.

Born and raised mainly in Ghana, but Dr Ayenor's connection to agriculture runs deep through Cuba and the Netherlands, where he studied and specialised. He grew up understanding both the potential and the challenges of African farming—lessons that have shaped his entire career. This personal connection to the land and to farming communities informs everything he does, from high-level policy work for governments and donors to grassroots farmer training.

A Career Dedicated to Agricultural Transformation

Dr Ayenor's career spans the full spectrum of agricultural and rural development. He has worked as a researcher, policy advisor, project manager, trainer, and entrepreneur—roles that have given him a uniquely comprehensive understanding of what it takes to create real change in African agriculture and rural settings.

As the Chief Executive Officer of STRED Consult Ltd. (Management Services), Dr Ayenor leads a consulting firm that has become synonymous with practical, results-oriented agricultural development across West Africa. STRED works with governments, international organisations, private sector companies, and farmer organisations to design and implement programmes that genuinely improve farmers' lives. The firm operates in both Ghana and Nigeria, giving Dr Ayenor direct experience with the opportunities and challenges of cross-border agricultural collaboration.

Through STRED, Dr Ayenor pioneered the STRED Digital Farming & Development Institute (SDFI), designed to be one of West Africa's leading initiatives to bring digital tools and modern agricultural knowledge to smallholder farmers. SDFI represents Dr Ayenor's conviction that technology—when properly adapted to African contexts—can democratise science and enhance access to information and markets in ways that transform what's possible for small-scale farmers and other actors, including petty traders, as a basis for industrialisation.

Shaping Agricultural Policy Across West Africa

Dr Ayenor has been instrumental in shaping agricultural policy at national and regional levels. He served as the principal author of Ghana's National Agricultural Policy (CPESDEP 2017-2024 and 2021-2025), a foundational document that continues to guide the country's agricultural development. He also contributed significantly to the development of Ghana's e-Agriculture Strategy, helping to chart the country's path toward digital transformation in agriculture.

His policy influence extends beyond Ghana. Dr Ayenor consulted for the ECOWAS MIRA Project (Mutual Recognition Agreement), which worked to harmonise seed and fertiliser regulations across West African countries. This work and another assignment on informal cross-border trade (UNECA – AfDB) gave him firsthand experience with both the potential and the challenges of regional integration—lessons that inform his thinking on AfCFTA and regional agricultural trade.

He has also contributed to the development of the Marshall Plan for Agricultural Transformation in Ghana, and a Cabinet paper on agricultural finance in Ghana. Always working to create comprehensive frameworks for sustainable agricultural development that balance productivity, economic prosperity, environmental sustainability, and social equity.

International Experience and Expertise

Dr Ayenor's expertise is recognised internationally. He has worked extensively with major development organisations, including the Food and Agriculture Organisation (FAO), International Labour Organisation (ILO), United States Department of Labour (USDOL), UNDP, WFP, the World Bank, the USAID, GIZ, British Aid, European Union (EU) and numerous other bilateral and multilateral agencies.

His work has taken him across West Africa and beyond, evaluating programmes, designing interventions, training practitioners, and advising governments. He has particular expertise in:

- Value chain development, especially for cocoa, shea, cashew, and other key African commodities

- Organic cocoa and certification systems, having worked extensively within the cocoa supply chain, including international cocoa marketing (demand side)

- Climate-smart agriculture and sustainable land management

- Digital agriculture and ICT for development

- Farmer organisation development and cooperative management

- Agricultural finance and market systems development (trainer in commodity structured trading systems)

- Gender integration in agricultural programmes

- Youth engagement in agriculture

- Project Management and Donor Project Evaluation

- Multi-Stakeholder Process Facilitator

A Practitioner Who Gets his Hands Dirty

What sets Dr Ayenor apart from many agricultural development professionals is his insistence on staying connected to the realities of farming communities. He doesn't just design programmes from offices in capital cities—he spends time in villages, on farms, in markets, and at border crossings, understanding what actually works and what doesn't.

He has trained thousands of farmers across West Africa in improved agricultural practices. He has facilitated the formation and strengthening of dozens of farmer organisations and facilitated connections in many commodity value chains. He has walked cocoa farms in Ghana's Eastern Region, visited cassava processing centres in Nigeria, and observed shea collection and processing in Northern Ghana. And stood at Ghana-Togo border posts and other borders watching trucks loaded with agricultural produce navigate bureaucratic obstacles.

This combination of high-level policy expertise and grassroots practical experience gives Dr Ayenor a perspective that few in the agricultural development field can match. He understands both what should happen in theory and what actually happens in practice—and, crucially, how to bridge the gap between the two.

Academic Credentials and Continuous Learning

Dr Ayenor holds advanced degrees in agriculture and development, with specialised training spanning value chain analysis to digital agriculture. But he's quick to point out that some of his most valuable education has come from farmers themselves—from observing what works in their fields, listening to their insights, and learning from their innovations.

He is a frequent speaker at agricultural conferences and workshops across Africa and internationally. He speaks Spanish in addition to English and has published numerous articles, reports, and policy papers on African agricultural development. But he's equally comfortable presenting to a room or under a tree full of smallholder farmers as to a conference of development professionals—and he considers both audiences equally important.

A Vision for Africa's Agricultural Future and its Environment

Throughout his career, Dr Ayenor has been driven by a single vision: African farmers prospering, rural communities thriving, and Africa feeding itself while contributing to global food security and protection of the environment for future generations. He has seen enough successes to know this vision is achievable, and enough failures to know it won't happen automatically.

His book *From Fields to Fortunes - How Africa Can Feed Itself and the World* " represents the distillation of everything Dr Ayenor has learned over three decades—the successes and the failures, the policies that worked and those that didn't, the innovations that *transformed* lives and the well-intentioned programmes that went nowhere. It's his attempt to share these lessons with the next generation of farmers, policymakers, entrepreneurs, and development practitioners who will shape Africa's agricultural future.

For more information about Dr Ayenor's work:

- **STRED Consult Management Services:** www.stredconsult.com
- **STRED Digital Farming and Development Institute:** https://stredconsult.com/about-sdfi/
- **Email:** info@stredconsult.com

Order the companion book at *https:// stredconsult.com/land-and-legacy-books/*

List of Acronyms

Acronym	Meaning
AfCFTA	African Continental Free Trade Area
AI	Artificial intelligence
ANSE	Neem extract (as referenced in the organic cocoa spraying context)
AU	African Union
B2B2F	Business-to-Business-to-Farmer (SDFI training model described in Chapter 7)
CAADP	Comprehensive African Agricultural Development Programme
CASA	Commercial Agriculture for Smallholders and Agribusiness
COCOBOD	Ghana Cocoa Board
CRIG	Cocoa Research Institute of Ghana
CSR	Corporate social responsibility
COVID-19	Coronavirus disease 2019
DVI	Digital Villages Initiative
ECOWAS	Economic Community of West African States
EU	European Union
EWG	Expert Working Group
FAO	Food and Agriculture Organization of the United Nations
FBO	Farmer-Based Organization
FoB	Free on Board
GIZ	Deutsche Gesellschaft für Internationale Zusammenarbeit (GIZ)
IFC	International Finance Corporation (World Bank Group)
ILO	International Labour Organization

Acronym	Meaning
LARC	Local Agricultural Research Committee
LBC	Licensed Buying Company
M-Farm	M-Farm (Kenya mobile-based market platform)
M-Pesa	M-Pesa (mobile money platform)
MIRA	ECOWAS Micro Reforms for African Agribusiness
MoFA	Ministry of Food and Agriculture
NGO	Non-governmental organization
P4P	Purchase for Progress
PPP	Public-private partnership
QR	Quick Response (QR code)
SDFI	STRED Digital Farming and Development Institute
STRED	STRED Consult Ltd (brand name as used throughout the book)
TIPCEE	Trade and Investment Program for Competitive Export Economy
TOFA	Traditional Organic Farmers Association
TVET	Technical and Vocational Education and Training
UNDP	United Nations Development Programme
USDOL	United States Department of Labor
USAID	United States Agency for International Development
WFP	World Food Programme

References

Akiwumi, P. (2022, September 30). Revitalising African agriculture: Time for bold action. United Nations Conference on Trade and Development. https://unctad.org/news/revitalizing-african-agriculture-time-for-bold-action

Ayenor, G. K. (2006). Capsid control for organic cocoa in Ghana: Results of participatory learning and action research (Doctoral dissertation, Wageningen University).

Ayenor, G. K. (2015). Assessment of environmental and socio-economic impacts of small-scale illegal surface mining in the cocoa landscape in Ghana (Report). United Nations Development Programme.

Ayenor, G. K., van Huis, A., Röling, N. G., Padi, B., & Obeng-Ofori, D. (2007b). Assessing the effectiveness of Local Agricultural Research Committee approach in diffusing sustainable cocoa production practices: The case of capsid control in Ghana. International Journal of Agricultural Sustainability, 5(1), 1–14. https://doi.org/10.1080/14735903.2007.9684814

Ayenor, G. K. (2021a). Independent interim evaluation of Adwuma Pa project [Evaluation report]. Sistema Familia Y Sociedad / CARE International. (Sponsored by U.S. Department of Labor, Bureau of International Labor Affairs)

Ayenor, G. K. (2021b). Author of agricultural and rural development contents. In Government of Ghana, Coordinated Programme of Economic and Social Development Policies 2021–2025. National Development Planning Commission. https://ndpc.gov.gh/media/CPESDP_2021-2025_21-11-22-2_FINAL_CORRECTED.pdf

Ayenor, G. K. (2023). Final evaluation of Adwuma Pa project [Evaluation report]. Sistema Familia Y Sociedad / CARE International. (Sponsored by U.S. Department of Labor, Bureau of International Labor Affairs). https://www.dol.gov/sites/dolgov/files/ILAB/evaluation_type/

final_evaluation/Adwuma-Pa-Ghana-Project-Final-Evaluation-Final-nonPII-508-compliant.pdf

Breard, P., & Ayenor, G. K. (2024). High-level evaluation of the ILO's Decent Work Country Programme strategies and activities in Western Africa, with emphasis on Côte d'Ivoire, Ghana, Nigeria and the Economic Community of West African States (ECOWAS) (2018–2023) [Evaluation report]. International Labour Organization. https://www.ilo.org/wcmsp5/groups/public/---ed_mas/---eval/documents/publication/wcms_hle_west_africa_2024.pdf

Ehui, S. K., & Odeh, K. (2025, May 20). Digital solutions in agriculture drive meaningful livelihood improvements for African smallholder farmers. Brookings Institution. https://www.brookings.edu/articles/digital-solutions-in-agriculture-drive-meaningful-livelihood-improvements-for-african-smallholder-farmers/

Essilfie, G. L., Baddoo, R. N. N., & Ayibor, S. (2025). Postharvest handling of indigenous vegetables in Ghana: Implications for reducing food loss and enhancing nutrition. Frontiers in Horticulture, 4, Article 1619846. https://doi.org/10.3389/fhort.2025.1619846

Food and Agriculture Organization of the United Nations. (2023). Pilot digital villages initiative in Africa: Findings of country assessments and recommendations for effective implementation. FAO. https://doi.org/10.4060/cc6131en

Fox, L., & Signé, L. (2022, June 28). Overcoming the barriers to technology adoption on African farms. Brookings Institution. https://www.brookings.edu/articles/overcoming-the-barriers-to-technology-adoption-on-african-farms/

Ghana News Agency. (2024, August 30). Post-harvest losses impoverishing Ghanaian farmers. https://gna.org.gh/2024/08/post-harvest-losses-impoverishing-ghanaian-farmers/

Government of Ghana. (2017). Coordinated Programme of Economic and Social Development Policies (2017–2024): An agenda for jobs; creating prosperity and equal opportunity for all (G. K. Ayenor, author of Agricultural and Rural Development sections). Ministry of Planning /

National Development Planning Commission. https://
www.mop.gov.gh/wp-content/uploads/2018/04/Coordinated-
Programme-Of-Economic-And-Social-Development-Policies.pdf

Hodder, G., Migwalla, B., & Pickup, S. (2023, July 12). Africa's agricultural
revolution: From self-sufficiency to global food powerhouse. White &
Case.

International Institute of Tropical Agriculture. (2016, May 10). IITA and NAQS
pilot digital Cassava Disease Surveillance platform. https://
www.iita.org/news-item/iita-and-naqs-pilot-digital-cassava-disease-
surveillance-platform/

National Development Planning Commission. (2022). Comprehensive Plan of
Action for the Elimination of the Worst Forms of Child Labour in Ghana
2021-2025. Government of Ghana. https://ndpc.gov.gh/media/
CPESDP_2021-2025_21-11-22-2_FINAL_CORRECTED.pdf

World Food Programme. (2008). Pilot project report: Smallholder value chain
development and local food procurement in Ghana (Swedish Grant
Project No. SWE/2007/001) [Internal project report]. WFP Ghana
Country Office.

Zegers, M. C. R., & Ayenor, G. K. (2021). Ending child labour and promoting
sustainable cocoa production in Côte d'Ivoire and Ghana (Final
report). European Commission. https://international-
partnerships.ec.europa.eu/system/files/2021-07/ending-child-labour-
cote-ivoire-ghana-2021-final-report_en.pdf

www.ingramcontent.com/pod-product-compliance
Lightning Source LLC
Chambersburg PA
CBHW012032140726
47990CB00009B/3198